COAL
TECHNOLOGY FOR BRITAIN'S FUTURE

COAL
TECHNOLOGY FOR BRITAIN'S FUTURE

M

© Macmillan London Limited 1976

All rights reserved. No part of this publication may be reproduced or transmitted, in any form or by any means, without permission.

Picture research by Jeanne Griffiths

Diagrams by Osborne Marks Associates and Berry/Fallon

ISBN 0 333 19418 7

Published 1976 by Macmillan London Limited
London and Basingstoke
Associated companies in New York,
Toronto, Dublin, Melbourne, Johannesburg and Delhi

Filmset in Great Britain by Servis Filmsetting Limited
Printed by Martin Cadbury Ltd, Worcester

British Library Cataloguing in Publication Data
Coal: technology for Britain's future
  Index
  ISBN 0-333-19418-7
  333.8's HD 9551.5
  Coal – Economic aspects – Great Britain

Contents

Foreword
Sir Derek Ezra   6

**1. The Significance of Coal
as a World Energy Resource**
Roger Vielvoye   8

2. Coal and Coalmining
Sir Andrew Bryan   29

3. The New Mechanisation
Professor E. L. J. Potts,
Dr R. K. Dunham and John Scott   52

4. Selby – The Mine
Michael Pollard   97

5. Selby – The Community
Jeremy Bugler   108

6. New Uses for Coal
W. T. Gunston   129

Index   142

Foreword

This book appears at a time when world attitudes towards energy have been transformed. Until 1973 energy, when it entered into calculations at all, was regarded as a stable resource of infinite capacity, something so permanent and constantly available that it could be safely left to take care of itself without causing concern to Governments.

The fuel crisis in October 1973 changed all that. Almost overnight, the action of the oil-producing countries in quadrupling oil prices and restricting output revolutionised the world situation. Energy became a major factor in global politics. For the first time it was widely realised that fossil fuels supplies were not infinite. Reserves of oil, even in the vast fields of the Middle East, would begin to dwindle after attaining peak production in the 1990s, and natural gas – already approaching the point of exhaustion in the United States – could not be seriously relied on when planning for the twenty-first century.

The one hydro-carbon fuel the world has in abundance is coal. There are large deposits in most continents, and although in Britain we have been working coal seams intensively for more than a century, we still have reserves which could be worked for at least another hundred years with our present machinery and equipment, and perhaps for many hundreds of years if new techniques can be developed for mining thin or hitherto inaccessible seams.

The British Government was among the first to see what the changed situation meant for coal. No longer could Britain rely on cheap imported oil while running down the mining industry. The decline in coal production which had been continuous since the mid-fifties had to be halted, and then reversed.

In a detailed examination conducted with the National Coal Board and the trade unions of the mining industry, the Government concluded that it was essential to the country's economy that coal should be revitalised, and gave full approval to a ten-year programme, 'Plan for Coal', that would double the historic rate of investment for the following decade and thereby restore the industry's production capacity.

The coal industry has made a good start on this plan. Investment is going according to schedule and the exploration programme to find fresh coal reserves has been establishing them at a rate four times greater than the current rate of mining.

The prospects for British coal are bright.

Our research and development programmes are forging ahead to improve the performance of mining machinery, using the most up-to-date methods of automatic and remote controls. Considerable resources are also being devoted to the processing of coal, and in conjunction with other coal mining industries and through the International Energy Agency, the N.C.B.'s Coal Research Establishment is working on the production of gaseous and liquid fuels as well as other sophisticated chemical derivatives from coal, and on the more efficient combustion of solid fuels using techniques such as fluidised bed combustion. The concept of the 'Coalplex' – in which heat, energy, and a multiplicity of by-products could be produced from a single coal processing 'refinery' – is one of the positive offshoots of the years of concentrated research by coal's scientists.

All this points to an expanding and prosperous future for the world's coal-mining countries. But prosperity will not fall into our laps: it will have to be planned and worked for. This book plots the paths which brought the coal industry in Britain to its present position. It also looks at the opportunities for the future. Choosing the right course and sticking to it will require all the technical and human skills at the disposal of Britain.

The prize for success will be a strong place among the economically significant nations of the future – the energy suppliers.

Sir Derek Ezra

1. The Significance of Coal as a World Energy Resource

Roger Vielvoye

In the two hundred years that have passed since the start of the industrial revolution, the countries of Europe and North America, together with Japan, have consumed coal, crude oil and natural gas that took over 200 million years to create. As living standards in these countries have improved, demand for energy has soared. In a modern industrial society committed to continuous economic expansion the growth of energy supplies has become as important as expanding food supplies. People in the developed countries expect to live in warm and comfortable houses during winter and cool homes during summer. They are surrounded by labour-saving gadgets in the home and have their tasks at work made less onerous by more complex machinery. And they take for granted the ability to drive freely in their own cars or travel abroad by aircraft. Every facet of modern life in the western world depends upon an adequate supply of energy.

But the demand for seemingly endless supplies of energy is strictly confined to the industrial countries. Fuel usage can be directly related to the economic strength of a country, and on this basis every man, woman and child in the United States uses the equivalent of ten tons of coal a year. Further down the economic league table, in Britain the consumption is five tons a year, and in Japan just over three tons. At the other end of the scale Indians consume the equivalent of well under a quarter of a ton of coal per head. It is estimated that 20 per cent of the world's population consumes over 85 per cent of its annual production of coal, crude oil and natural gas. Underdeveloped countries, on the other hand, are still heavily dependent on what are known as 'non-commercial' fuels such as wood, dung, and vegetable and other wastes. The United Nations estimated that in the early 1950s about 15 per cent of the non-communist world's energy requirements were obtained from these materials. By 1967 only 4 per cent of the total world energy needs were derived from 'non-commercial' sources, yet nearly half the world's population was dependent on them.

With the world using up its reserves of fossil fuels a million times faster than they were created the spectre has reappeared of a world-wide shortage of energy – in twenty-five years' time crude oil and natural gas could be in short supply unless consumers are prepared to develop and use other forms of energy. On the surface it might appear that altering the energy base of a complex economy would be a difficult and lengthy task. Yet in the past fundamental changes of this sort have been made remarkably smoothly. After the Second World War, Europe changed from being a coal-burning continent to one heavily dependent on cheap oil almost exclusively imported from the Middle East and Africa. During this period cheap oil also transformed the Japanese energy economy and turned the United States from being self-sufficient in energy into a net importer of oil. But in each case the transition took place painlessly, mainly because the new fuel – oil – was cheap, easy to handle, and (at the time) supplies seemed almost unlimited.

But by the early 1970s it was clear that the oil companies were finding it difficult to maintain the prodigious rate of new discoveries that in the past had ensured that oil was plentiful and therefore cheap; while coal reserves, for instance, in Britain were (and are) being proved at a rate four times greater than they were being mined. The basic concept of cheap oil that made it difficult for coal as well as other forms of alternative energy to compete in the market place was questioned, and once again the providers of energy began to look at ways in which a new transition could be made from heavy dependence on oil into a multi-fuelled world energy economy. International oil companies began to turn themselves into energy corporations. They bought interests in coal, nuclear fuel, nuclear reactor construction and other alternative energy sources.

Whether the industry could have made another peaceful transition is now largely hypothetical. In October 1973, the Arabs and the Israelis yet again went to war and the Arab oil-producing nations deployed the famous 'oil weapon'. At the same time, but in a different forum, the Organisation of Petroleum Exporting Countries (OPEC), the oil producers, disbarred the major international oil companies from having any future role in deciding the world oil pricing structure. The result of these two simultaneous and parallel moves is now well known. The Arabs banned all exports of oil to the United States and Holland and reduced deliveries to many of the other leading Western nations. Shortages occurred almost immediately and the value of oil left in world trade soared immediately: within three months it had increased fourfold.

Consumers groaned. They could not get enough oil to meet their demands and all sorts of rationing schemes were introduced. Sunday driving was banned in a number of countries and strict speed limits were imposed. Not only was oil in short supply but the barrels that were available cost four times as much. Overnight, energy users in the industrialised countries were given a taste of life in a real fuel shortage. Converts to the previously barely recognised conservationist camp came in their thousands and governments began to take very seriously the business of planning their future energy supplies. Coal, nuclear power and more futuristic forms of energy production now came into their own as paying propositions. The governments of all the major energy-consuming countries, with the exception of France, banded together under the banner of the International Energy Agency to develop an emergency oil-sharing system that would resist the threat of a future curtailment of supplies from the OPEC countries. It was not purely a defensive organisation. The I.E.A. began to look seriously at joint projects to finance research into alternative fuels, including coal. One thing had been learned during the 1973–74 supply and price crisis – countries could no longer pursue nationalistic energy policies or allow the energy-producing industries to compete with each other regardless of the consequences.

The curtailment of supplies of oil at the same time as the quadrupling of the price emphasised how deep-rooted in everyday life was the assumption of the availability and low cost of energy. Higher energy prices have since increased the cost of producing many goods, increased fares on public transport, made private motoring more expensive, and are already forcing people in the industrial countries to adopt a new attitude to conservation in home heating and communications. Governments and industry are engaged in detailed energy audits to discover just where fuel is being used most inefficiently and have spent large sums of money on encouraging people to save energy.

Energy crises, however, are nothing new. Ever since man started chopping down trees and burning wood, prophets of doom have been warning of impending disaster unless new methods of providing power and heat were discovered. In 1865, an economist named Stanley Jevons published a book which forecast a serious coal shortage. His book triggered two Royal Commissions, one of which was pessimistic about coal's future prospects. Similar forecasts of shortages have come at regular intervals in the oil industry but each has been followed by the discovery of new reserves. The latest forecasts of oil running out by the turn of the century will need the discovery of new reserves on a scale not previously seen. An area with the potential of the North Sea will have to be found every two years, a schedule the industry has failed to maintain during the 1970s because the most likely oil prospects left for drilling are in deep offshore waters. Although the experts are pessimistic about uncovering a new oil-bearing region to rival the Middle East, some of the leading oil producers, mainly Saudi Arabia, are hinting that still uncharted reserves may yet be available as a result of future exploration programmes.

In these circumstances it would be easy to dismiss the latest forecasts of coming energy shortages as alarmist and rely on technology once again to produce a solution. However, even the technologists are more cautious about dismissing the threat of rapidly declining fossil fuel reserves out of hand. There is general agreement, in fact, that at current rates of consumption the world's oil and natural gas reserves could be largely exhausted by the turn of the century. The most pessimistic of these estimates came in 1972 from the Club of Rome in their report *Limits to Growth*. By extrapolating rates of population growth and energy consumption into the future, it concluded that oil reserves would be exhausted within twenty years; natural gas in twenty-two years; and coal in 110 years. Meanwhile a sixfold increase in oil prices since the end of 1973, combined with a serious world economic recession in 1974 and 1975, has temporarily checked the rate at which energy demand is growing. Nevertheless, supporters of the view that the link between energy use and population growth is vital in considering the future prospects for energy feel that, while it is probably over-pessimistic, the premise of the Club of Rome still holds good.

Followers of this neo-Malthusian view, that global catastrophe is imminent unless zero population growth is achieved within the next twenty years, see developments in mining techniques, nuclear power, the greater use of more plentiful coal reserves, and other alternative sources of energy merely as delaying measures that should take second place to tackling the real problem of overpopulation.

Against this 'technological optimists' (as they have become known) show an impressive range of developments in energy production, the exploitation of which over the next twenty to thirty years could, they feel, largely solve the world's power problems. In the immediate future they see the development of coal through new mining techniques and the opening of previously unexploited reserves as the best answer to the political uncertainty, rising price and declining reserves of oil. Improvements in the performance of nuclear power could make a significant contribution. Oil could also be made available from the massive shale reserves in the United States and from the tar sands in the Lake Athabasca area of Canada and from Venezuela. Delving further into the future there is the development of solar power and the harnessing of the winds and waves to look forward to, together with the commercial exploitation first of the nuclear-breeder reactor, and then of the successful production of large quantities of energy through the nuclear-fusion process.

On top of this the technologists are continually working on the problem of reducing the wasteful and profligate way in which energy is used and trying to increase the amount of energy produced from a given amount of fuel. Largely as a result of increased efficiency between 1920 and 1965 fuel requirements in the United States fell from 149.1 million to 91.1 million joules per dollar of the country's Gross National Product.

But, as the Malthusians in the energy debate will eagerly point out, between 1965 and 1970 the trend was reversed and in five years the number of joules per dollar of G.N.P. rose to 100.2 million. If this trend continues the case of the technological optimists for ever increasing efficiency in the use of energy must be undermined.

The case for population control is not seriously disputed, but it should be pointed out that in the present state of the energy market, even very large changes in population growth in the developing countries would have an almost negligible effect on energy demand, while relatively small changes in the United States or Europe could have quite startling effects on the amount of energy needed to keep society functioning at present standards.

Of all the options open for filling the energy gap, the greater exploitation of coal could be accomplished with the least expenditure on research and development, although vast sums would be needed to

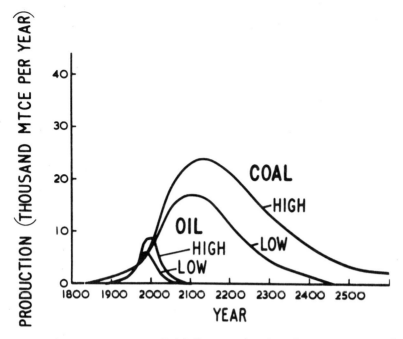

A projection (by M. K. Hubbert) of coal and oil production, measured in million tons coal equivalent.

open up the new reserves. Initially, production from an expanding coal industry would be channelled into electricity generation. Here nuclear power also has a great role to play, but unlike coal, there are many technological problems still to be solved in the quest for cheap and plentiful supplies of electricity from the atom, and the world research budget must be substantially increased if the targets set for nuclear power are to be achieved.

Coal's great strength lies in the extent of its reserves and the fact that most of these can be opened up *without* the aid of new technology. Currently it is estimated that there are over 6,700 thousand million tons of coal reserves throughout the world. While the Club of Rome takes a pessimistic view of coal and puts its useful lifespan at 111 years, the most optimistic view, based on the ability of the mining industry to extract 50 per cent of these reserves, gives coal the ability to meet demand for another 400 years.

Reverting to coal as a major world energy source would be a return to a trend started 2,000 years ago by the Chinese. The use of coal instead of wood as a source of heat originated in ancient China (although there is evidence that the Romans were also aware of the properties of coal). During the Dark Ages the little use that had been made of coal for fuel in the home was abandoned in favour of wood, beginning serious deforestation on the plains of Northern Europe, a trend that was already under way in China.

Between 1000 and 1200 AD the first attempts were made to establish a mining industry in England, Scotland and parts of continental Europe. Methods of extracting coal were crude (see Chapter 2). It was hewn from outcrops on the surface or hauled out of shallow pits, a few metres deep, sunk into surface seams. Despite the unsophisticated methods, demand for coal grew as a replacement for charcoal for ironmaking, and the brick and glassmaking industries. By 1550 Britain had reached a pre-eminent position in the mining world and was producing over 200,000 tons of coal a year.

It was during this period that Britain underwent its first energy crisis. Wood was still the primary source of energy and in such short supply that in 1593 the price of firewood rose by over 800 per cent, making even the present-day increases of oil prices by OPEC seem modest. Eventually the British Parliament was forced to take action. Exporters were required to return empty beer and wine casks or import foreign lumber to replace the supply.

Mining became more important over the next two hundred years as wood became more and more difficult to obtain for fuel. So by the time Britain and other continental countries began to develop a real industrial base in the eighteenth century most of the surface seams were worked out and coal owners were forced to begin much deeper underground activities. M. K. Hubbert in an article on energy resources estimated that world consumption of solid fuels increased at a rate of two per cent between the year 1000 and 1860. During this time up to 7,000 megatonnes were consumed. World consumption from 1860 to 1914 rose at an annual rate of 4.4 per cent, but then the growth slackened to only 0.75 per cent a year between 1915 and 1945. From the end of the Second World War the growth rate has averaged 3.6 per cent a year.

The development of coal in the United States followed a similar pattern to that of Europe. A coal industry grew up on the east coast often as deforestation took place. As in Europe coal reigned supreme as an energy source throughout the nineteenth century and up to the outbreak of the First World War. However, from 1859 the oil industry began to emerge.

Like coal, it had been known as a source of fuel from the earliest times. The Chinese and peoples in the Middle East used crude oil seeping up through the ground naturally to burn or, in its more bituminous form, as a waterproof lining for their boats. In Burma came the idea of digging shallow pits close to natural seepages and allowing them to fill with oil. Then in 1859 Col. Edwin Drake brought the techniques developed from sinking water wells into the oil search – and the modern oil industry had begun.

Initially the new oil industry and the established coal interests were not in competition. Oil was used for fuelling kerosene oil lamps while coal maintained its stranglehold on industry, on the railways, in the shipping industry and as a source of home heating. In 1900 the international consumption of coal was 725 million tons, or 93 per cent of total energy consumption. Before the outbreak of the First World War, oil began to replace coal in the navies of the world; and then, between the wars, began the age of the automobile. By 1940 oil had made considerable inroads into coal's dominance. Demand for coal was then running at 1,500 million tons a year – 71.4 per cent of total consumption – while the oil companies were selling 600 million tons.

The real change in the world's energy consumption patterns started

Overleaf: A map of the world's coal resources.

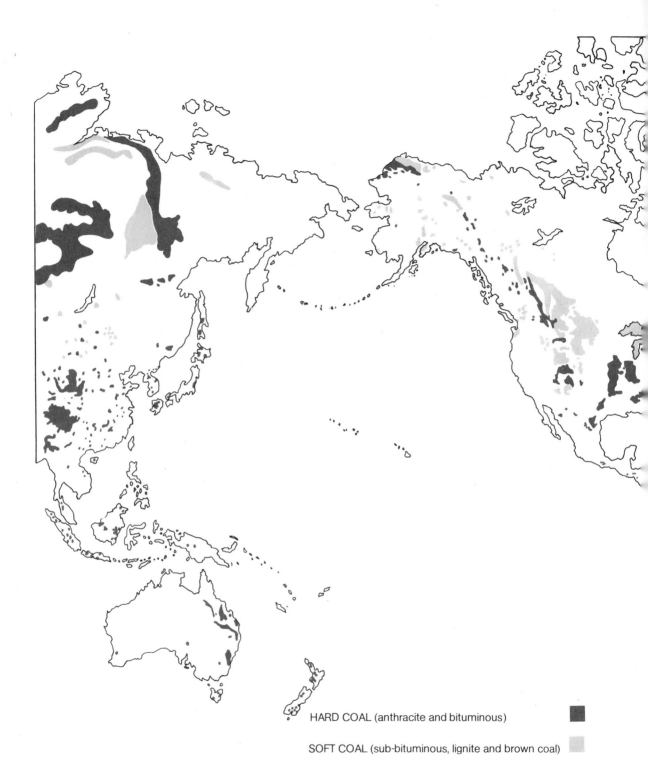

WORLD COAL RESOURCES

after the Second World War. New oilfields discovered in the Middle East needed additional market outlets and the oil companies began an aggressive campaign to sell their wares in Europe in the same way as they did in the United States. Oil was introduced for steam raising in power stations, previously a preserve of the coal industry, and it began to replace coal in other areas of industry. By 1950 coal output was up to 1,550 million tons a year, but as oil demand was up to the equivalent of 1,150 million tons a year coal's share of the market had dropped to 67.4 per cent. In the following decade oil continued to whittle away at coal's share, but massive increases in the demand for energy allowed this change of balance to take place without the overall output of coal falling dramatically. In fact by 1960 output of coal had risen to 2,100 million tons a year to share the market equally with oil and gas. It was in the 1960s that oil consumption really took off. In the twelve years between 1960 and 1972 consumption of oil and gas soared to the equivalent of 5,300 million tons of coal, giving the oil companies about 68 per cent of the total market.

While the rate of growth in consumption has been slow during the period in which oil has become the world's dominant fuel, the process has brought a significant change in the principal suppliers of coal. According to the World Energy Conference survey of energy resources published in 1974, about two-thirds of all solid fuels ever consumed have been mined in the period from 1925 to 1974. In the pre-1925 period three nations – the United States, Germany and the United Kingdom – accounted for 80 per cent of all coal and lignite mined to that date. Statistics for the year 1925 show that the U.S.A., Germany and the United Kingdom accounted for 76 per cent of world production and were the only nations producing over 100 million tons of coal a year. In 1925 an additional fourteen nations, half of which were in Europe, were producing more than ten million tons a year.

In 1971, the situation had changed completely. Total world output had risen threefold. Output in Britain and America was reduced. East and West Germany had improved their output but four other countries were now producing over 100 million tons a year. The significance, however, of the change in patterns of production on world energy markets is diminished as the four newcomers to the league of those producing 100 million tons a year or more are all from the communist bloc. The Soviet Union, from producing only 25 million tons a year in 1925, is now the biggest producer in the world with over 620 million tons annually. China, from 24 million tons in 1925, now has an output of 410 million tons annually. Two other east European countries – Poland with an annual output of 180 million tons and Czechoslovakia with 113 million tons – are also among the largest producers. But the communist bloc has in the past played very little part in influencing trends in the energy market. Poland exports important amounts of coal and the Soviet Union has made occasional forays into the oil exporting business.

But an analysis of the figures does show that in the areas where the oil industry made its greatest efforts to sell the increasing output of crude oil from the Middle East and North and West Africa, the indigenous coal industry has suffered. Britain, which was producing 247 million tons of coal a year in 1925, was down to 147 million tons a year in 1971 after an extensive programme of colliery closures. France's coal industry had also contracted sharply and output in Belgium was less than half what it was in 1925. By contrast, as well as the huge increases in the eastern bloc countries, there was a rapid rise in production in both the developing countries and in areas where there were plentiful supplies available at low cost. Australia and South Africa were now major producers and India built up its coal industry to the point where output was over 70 million tons a year.

Coal production should also have suffered from the advent of cheap electricity from nuclear power stations. But after the early optimism of the late 1950s, nuclear power has failed to live up to its initial promise. Britain's domestic construction programme has been expensive, dogged by technical problems and has slipped years behind schedule, but to its credit has produced power as cheaply (if estimated by operating cost, excluding capital investment) as the most efficient oil- and coal-fired power stations without raising any of the serious safety worries of the American reactor-building programme.

The American nuclear programme is based on the scaling up and development of the water-cooled reactor systems originally developed for the American nuclear submarine programme. Safety fears have made it difficult for American utilities to get licences to build nuclear stations in the United States, and it probably takes as long to get permission for an atomic reactor station as it does to build it. Outside America and Britain, the Soviet Union and Canada (which has been spectacularly successful in developing its own heavy-water reactor system), national reactor developments have generally been unsuccessful and have given way to local licensing of the two American systems available.

Despite the setbacks the technical optimists still have faith in the ability of nuclear reactor builders to solve the problems that have dogged the industry, providing sufficient finance is made available. They also set great store by the development work being done on the breeder reactors now under development in the United States, Britain, France and the Soviet Union. Building a nuclear reactor that actually 'breeds' nuclear fuel while producing electricity is a technical dream that is approaching reality. Britain, France and the Soviet Union all have working prototypes and are gearing up to construct the first commercial system. Problems, however, abound in scaling up this most complex piece of technology and, as has been proved in the past, the transition can be long and painful.

A new constraint on the spread of nuclear technology in the developing countries for producing electricity has arisen with India's explosion of a nuclear weapon using technology and fissile material supplied by the Canadians, under a technology deal in fact intended to help India produce cheap electricity. The nuclear powers are becoming reluctant to sell their data on peaceful uses of nuclear power to countries that might have an interest in using it to build up their capacity to construct nuclear weapons.

The European Community's energy planners see nuclear power as the principal way of reducing Europe's overall dependence on imported oil. They have recommended that nuclear power stations producing as much power as conventional stations burning 260 million tons of oil should be built by 1985. This would represent about 16 per cent of the Community's energy requirements instead of the 2 per cent currently supplied by atomic power. So far, however, none of the members of the community, with the possible exception of France, has a nuclear power station building programme geared to this rapid level of expansion. Britain, with North Sea oil available and ample reserves of coal, has criticised the programme as too ambitious, and even the Germans, who are desperately anxious to ease the stranglehold oil has on their own energy economy, do not feel the planners are being entirely realistic.

In fact, the world resources of solid fuel are enormous. The World Energy Conference's 1974 survey of global energy resources, the most thorough study we have to date, showed there were 10.8 million megatonnes of solid fossil fuel reserves. Ninety per cent of these resources are concentrated in three countries, the United States, the Soviet Union and China. Figures of this size are almost too large to comprehend, but when it comes to deciding how much of these resources is likely to be economically exploitable, then the numbers are drastically reduced. Only 1.4 million megatonnes are thought to be commercially exploitable at some time in the future, and of these only 0.6 million megatonnes could be recovered at current prices and using present-day technology. As can be seen from the chart, continually rising energy prices, combined with new extractive technology and developments in the use of coal as a chemical feedstock or liquid fuel form, would enable the amount of recoverable reserves to be increased sharply. Statistics on coal reserves are hampered by the fact that little new exploration work has been done in many areas, particularly Europe and North America, where the industry has been static or declining over the past twenty years in the face of competition from oil.

Resources as at 1974 by Continents, and Nations with Major Resources (all quantities in megatonnes)

Country or Continent	Reserves		Total Resources
	Recoverable	Total	
USSR	136,600	273,200	5,713,600
China, P.R. of	80,000	300,000	1,000,000
Rest of Asia	17,549	40,479	108,053
United States	181,781	363,562	2,924,503
Canada	5,537	9,034	108,777
Latin America	2,803	9,201	32,928
Europe	126,775	319,807	607,521

Africa	15,628	30,291	58,844
Oceania	24,518	74,699	199,654
World Total	591,191	1,402,274	10,753,880

Coal Reserves and Resources in Europe as at 1974
(all quantities in megatonnes)

Nation	Reserves		Total Resources
	Recoverable	Total	
Western Europe			
West Germany	39,571	99,520	286,150
Netherlands	1,840	3,705	3,705
France	458	1,407	1,407
Belgium	127	253	253
Austria	64	148	177
Total	42,060	105,033	291,692
Southern Europe			
Yugoslavia	16,870	17,976	21,751
Spain	1,643	2,202	3,562
Greece	680	908	1,575
Italy	33	110	110
Portugal	33	42	42
Total	19,259	21,238	27,040
Northern Europe			
United Kingdom	3,871*	98,877	162,814
Finland (Peat)	4,290	33,000	33,000
Sweden (Peat)	30	9,460	9,490
Iceland (Peat)		2,000	2,000
Denmark (Peat)	20	561	581
Ireland (Peat)	418	422	448
Norway	2	2	152
Total	8,631	144,322	208,485

* This U.K. figure was calculated on a different basis from those of the other countries. For the 1977 World Energy Conference it has been revised to 45,000 megatonnes.

Eastern Europe			
Poland	22,640	38,874	60,600
East Germany	25,300	30,200	30,050

Czechoslovakia	6,363	13,774	21,430
Hungary	1,675	3,350	6,400
Bulgaria	4,387	4,387	5,230
Romania	1,150	3,970	1,960
Total	61,515	94,555	125,670
Total Europe	131,465	365,148	652,887

Most of the European countries outside Scandinavia have coal resources of some sort. The Scandinavians make up for their lack of coal with larger quantities of peat, a source of fuel that these countries are taking more seriously in the light of current oil prices and long-term uncertainties about the supply of other fossil fuels. In Western continental Europe, 98 per cent of coal resources are concentrated in West Germany. Although in percentage terms many of the others like France and Holland have run down their indigenous mining, and have relatively small recoverable reserves left, these could still make a useful contribution to meeting European requirements. Only 4 per cent of Britain's total solid fuel resources are currently recoverable but the National Coal Board's dramatically increased exploration programme should provide sufficient new data on underground reserves to enable these figures to be updated.

The European Economic Community has been forced to revise its plans for coal in meeting its energy requirements and lessening dependence on imported oil. In 1974, coal and lignite accounted for 24 per cent of the Common Market's primary energy requirements, about four-fifths of which were from indigenous production, the balance being imported from outside, mainly from Poland and the United States. The European Commission has decreed that coal's share of the EEC energy market shall not fall below 17 per cent by 1985, although before the 1973/74 crisis it had been estimated that coal's share in ten years time would have declined to just 10 per cent. While the percentage total satisfied by coal will have dropped, the overall increase in energy consumption over the next ten years will mean that coal consumption in the EEC must be about 300 million tonnes by 1985 compared with the 281 million tonnes consumed in 1973. In a situation where the European coal industry has suffered from insufficient new investment for nearly two decades, substantial sums will have to be spent in new production capacity just to replace pits that are coming to the end of their reserves. In these circumstances most coal utilities feel that adding an additional 20 million tonnes of new capacity is hardly feasible and that this gap will be filled by further imports.

Britain, through the National Coal Board, does not entirely agree with some of the continental producers' estimates of the likely volume of imports from outside the community. It feels that its ambitious ten-year development programme (aimed at providing an additional 42 million tons of deep-mined capacity and an extra 5 million tons a year from opencast working) can ensure that Britain can reduce the com-

A map of the National Coal Board areas and coal fields.

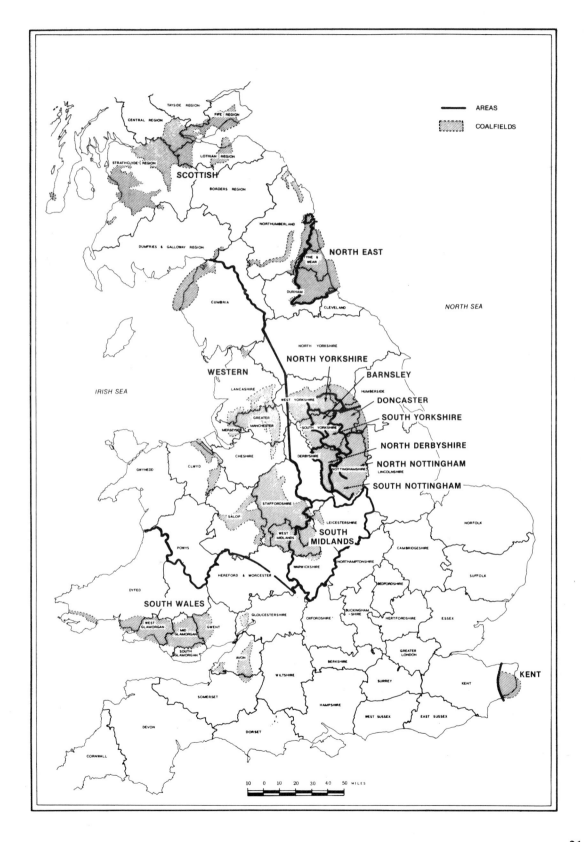

munity's call on coal from outside sources. This programme will enable the N.C.B. to maintain and expand its current production capacity of 114 million tons of deep-mined coal and 10 million tons of opencast as pits that have been in production for many years exhaust their workable reserves. The extra production capacity will enable the United Kingdom to become once again a more significant force as an exporter of coal.

Centre piece of the N.C.B.'s development strategy is the new £400m complex at Selby in Yorkshire designed to produce up to ten million tons of coal a year by the mid-1980s. Work has also begun on sinking a new anthracite mine at Betws in South Wales and on a new drift mine at Royston in Yorkshire: each of these projects will produce about half a million tons of coal a year. Twelve million tons are to be extracted over the next ten years from the vast opencast site at Butterwell in Northumberland. In addition existing mines will be expanded and new seams tapped. By the end of 1975 the N.C.B. had approved fifty-eight of these internal expansion projects, which should provide an additional capacity of over 10 million tons annually. They will shortly be able to decide the sites for all the new mines up to 1985.

Planned expansion has led to the biggest coal exploration programme in Britain for many years. The N.C.B. knows the rough location of much of the country's solid fuel resources but has lacked precise information on the thickness and depth of seams. The exploration programme is designed to provide this detailed information so the N.C.B. can decide which seams are commercially exploitable and where new pit workings should be sited. During the 1975/76 financial year the N.C.B. drilled 102 deep boreholes and is adding new proven reserves at the rate of 500 million tons a year, nearly four times as fast as coal is being mined. In fact, there were more rigs drilling for coal in that period than there were in the North Sea looking for oil.

Britain's coal industry has been able to remain relatively healthy when compared with most continental producers, because the British electricity generating network remained heavily dependent on coal when other power utilities made a rapid change to oil to take advantage of cheap prices. Over 60 per cent of Britain's electricity is generated by coal; in Europe the percentage is only about 30 per cent.

After leading the world in the commercial uses of nuclear power, the percentage of Britain's electricity coming from this source has declined. Until the beginning of 1976 nine nuclear stations accounted for about 5 per cent of the country's available generating plant. Of these, eight were in England and Wales and their operator, the Central Electricity Generating Board found them £84m a year cheaper to operate than fossil-fuelled stations. In early 1976 the first of the advanced gas-cooled reactors made their initial contribution to the nation's electricity grid. The construction programme for this type of reactor is running at least four years behind schedule and is likely to cost nearly £500m more than originally estimated – mainly because of technical problems and changes in the overall design.

The desperate troubles of the advanced gas-cooled reactor pro-

Right: Betws drift, a new mine in South Wales expected to start producing valuable anthracite in 1978.
Below: A general view of Longannet mine, Kincardine looking south. Here a mine complex has been successfully created and blended into the local landscape.

Overleaf: The total United Kingdom consumption of primary fuels in 1964 and 1974 (left); and the comparative input of primary energy to the electricity and gas industries in the United Kingdom between the years 1960 and 1974 (right).

TOTAL INLAND CONSUMPTION OF PRIMARY FUELS 1964 & 1974

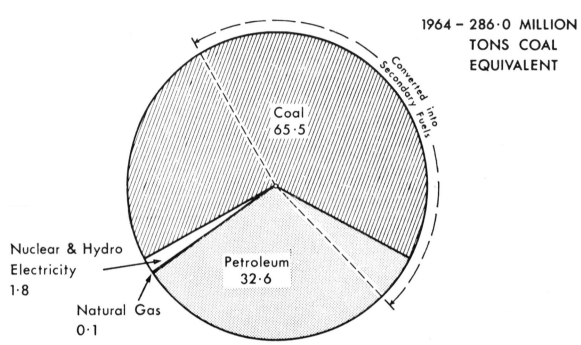

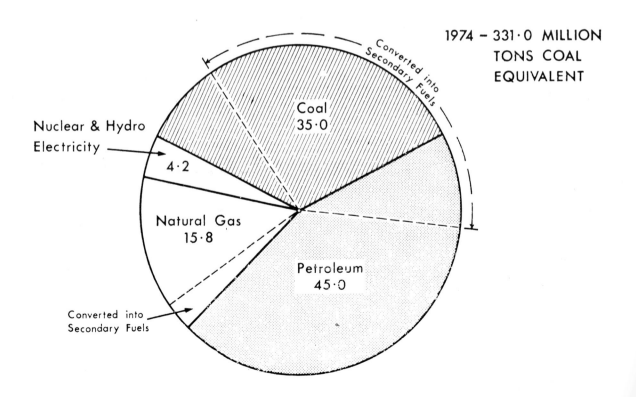

INPUT OF PRIMARY ENERGY TO THE ELECTRICITY & GAS INDUSTRIES

MILLION TONS COAL EQUIVALENT

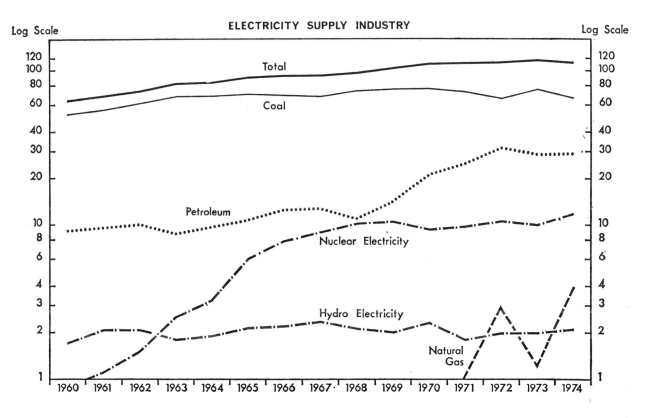

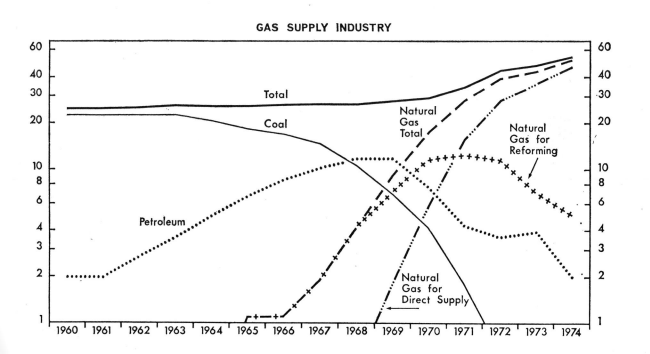

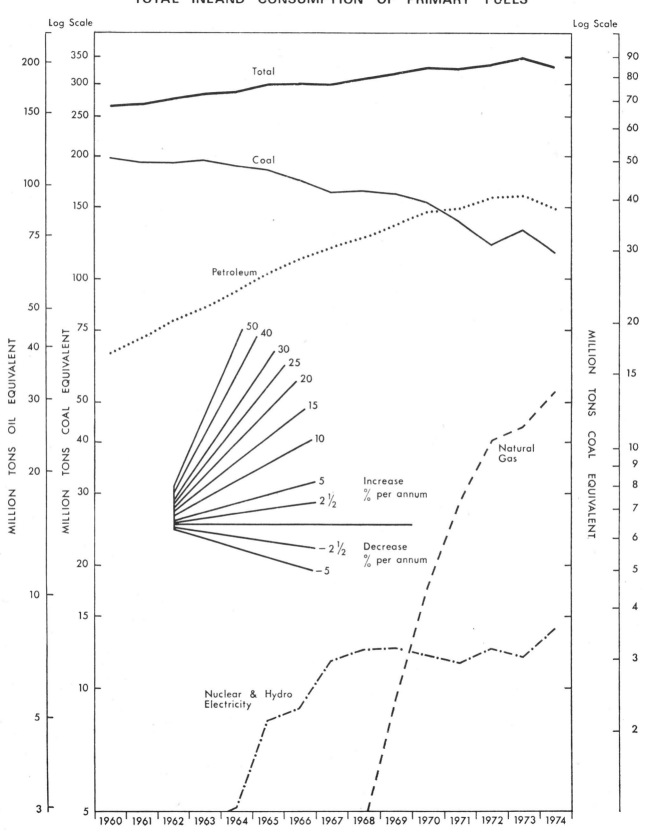

Left: The scale of primary fuel consumption in the United Kingdom between 1960 and 1974.

gramme led to a reorganisation of the reactor construction industry in 1974, and the emergence of a single company, the National Nuclear Corporation. After much debate, the Government ignored the case for building American-style light water reactors and told the new company it would have to build two large nuclear stations using the all-British steam generating heavy water reactor system.

Government long-term plans make provision for coal and nuclear power sharing most of the increased energy demand from the 1980s. There are no guarantees that the steam generating heavy-water reactor is anything more than a stopgap reactor (although a report in early 1976 by the Organisation for Economic Cooperation and Development's Nuclear Energy Agency estimated that by the turn of the century there would be 77,000 megawatts of SGHWR capacity in use throughout the world): Britain is also banking heavily on the success of the fast breeder reactor. This is the nuclear system that actually produces more nuclear fuel than it consumes. With questions being raised over the long-term security of uranium supplies, it is plainly vital that this generation of reactors is developed commercially. (A prototype has been built at Dounreay in Scotland.)

Looking outside Britain, 13 per cent of the total area of the United States, including Alaska, contains (as estimated in 1974) coal bearing strata. The largest reserves are in the western Great Plains and Rocky Mountain areas, which account for about 66 per cent of the country's total fuel resources. Substantial amounts of coal are to be found in the Appalachian basin. There the industry is well established and data on the geology is better known than in the western areas. Despite the size and easy access to a large proportion of this coal, the American industry has suffered since the mid-1950s from competition from cheap oil and even cheaper natural gas. Now that the country's self-sufficiency in energy has disappeared and the vulnerability of the economy to an interruption in imported oil supplies demonstrated, successive administrations in Washington have sought to revive coal production and so reduce the volume of oil imports. President Ford has set a target of doubling the 1973 output level of 590 million tons by 1985.

Over 50 per cent of the current U.S. output comes from opencast or strip mining on the surface, and providing permission can be gained there are ample opportunities for increasing the amount of coal taken out by this method. The World Energy Conference estimated that 5 per cent of all coal resources in the U.S. were at depths of 10 metres or less and probably 15 per cent are less than 100 metres deep. The Belle Ayr South mine in Wyoming, for example, had a capacity of 4.5 million tons in only its second year of surface mining operation in 1975. Development of the companion Belle Ayr North mine will give an estimated 25 million tons of output annually by 1978 and 40 million tons a year by 1984. Such a massive output is possible because the seam of coal is an incredible 21 metres (70 feet) thick.

Yet, because of its strip mining activities in the past, the American mining industry is now running into environmental objections to its expansion plans. For many years strip mining took place without any thought for replacement of the surface: with the result that many areas are still badly scarred.

Information about the reserves in the U.S.S.R. and China is more scanty, and neither of them is likely to make an impact in the small but growing world trade in coal. Russia has enormous resources in Siberia and its near-polar areas but is more interested in developing smaller deposits located closer to centres of consumption. In China, coal is still the major source of energy, although attempts are being made to develop its ample oil reserves rapidly. India is the only other Asian country with a thriving coal industry and reserves capable of sustaining long-term production. In the southern hemisphere, two major producers, Australia and South Africa, have easily exploitable resources and have plans to increase exports of cheap coal.

Against this international background coal is growing in stature. It forms the backbone of the International Energy Agency's plans to develop alternatives to imported oil. Britain, Germany and the United States are cooperating on an exciting project to develop a system for burning coal more efficiently. The technique is known as fluidised bed combustion (see Chapter 6) and a prototype plant is under construction in Yorkshire. The Agency has also established a data bank on world coal reserves and resources with the aim of producing a broadly co-ordinated programme for exploiting coal fields.

But most important, from the coal producers' viewpoint, the eighteen members of the International Energy Agency have agreed to a minimum safeguard price for oil of $7 a barrel. Each member agrees that it will not sell imported oil at less than $7 a barrel on its home market, even if it is purchased for less outside.

In theory, without this agreement the oil producing nations who are extracting oil from the ground for less than a dollar a barrel in many cases still have the ability to undercut the price of fuel from new energy developments if they look like affecting oil's share of world markets. In practice, few if any of the oil producers would want to undermine the oil-pricing structure they have fought so hard to establish, ensuring world coal producers of a bright future for years to come.

2. Coal and Coalmining

Sir Andrew Bryan

There is no record of when, where and how coal was discovered. Tradition tells us that early man first found it by accident when he noticed that in some of the places where he built his wood fires the very stones in the ground on which the fires were built also began to burn. Five centuries ago, Marco Polo recorded that some thousands of years earlier the Chinese had already recognised 'a black stone that burns'. Doubtless coal was first used as a source of warmth for human comfort; but it may have been used in other ways. Analysis of certain ash remains in South Wales suggests that coal had been used for cremation in the Bronze Age (c. 3000 BC). There is evidence, too, that varieties of coal were used in ancient times to make ornaments. But it was probably not until the Roman occupation that it was known and used on any appreciable scale in Britain.

The occurrence of coal and its vegetable origin
Coal is a sedimentary rock. It occurs as a sequence of layers or seams separated by layers of other rocks, mainly shale, sandstone and fireclay. The main coalfields of Britain are found in two groups of sedimentary rocks, the Coal Measures and the Carboniferous Limestone Series. These were laid down in the Carboniferous (or coal bearing) Period, which began more than 300 million years ago and lasted about 75 million years. With the exception of the Thick Coal of Staffordshire and Warwickshire which is between 8 and 10 metres thick (26 and 32 feet), they contain as many as twenty or thirty seams, varying in thickness from 20 centimetres (8 in.) to 3 metres (10 feet) each and extending over a wide area. The average thickness of the seams that are worked is 1.5 metres (5 feet).

Because the imprints of plant leaves and stems are often found in the strata forming the immediate roof of coal seams, and the remains of tree and plant roots in the rocks forming the seam floor, the origin of coal had long been suspected to be vegetable. Modern chemical analysis shows that it contains much organic material, a feature that makes coal unique among sedimentary rocks. The variable proportion of inorganic material always present in coal is in admixture as an alien ingredient, and not in chemical combination with the organic material. Under a microscope coal shows the presence of plant fragments; examination of thin slices reveals traces of an original vegetable structure. So the evidence points conclusively to the vegetable origin of coal.

An artist's impression of swamp conditions in Carboniferous times.

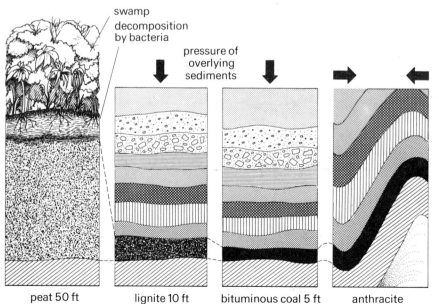

Left: A diagram showing the process of decomposition, subsidence, and pressure by which the progressive ranks of coal are formed, anthracite being the highest of all.

The formation of the sequence of coal seams found in the Carboniferous rocks in Britain extended over many millions of years, when practically the whole of the area on which Britain now stands was largely a vast shallow estuary or lagoon into which great rivers flowed. The area was slowly sinking at such a rate that, for most of the time, the sand, mud and clay deposited by the rivers on the bottom of the estuary kept the depth of the water fairly constant. At long intervals a decline in the rate of sinking occurred so that the silt nearly filled up the shallow estuaries, turning them into great swamps in which 'swamp' trees, giant ferns and other vegetation quickly grew. Encouraged by the warm, monsoon-type humid climate of Carboniferous times dense forests were formed. The swamp conditions persisted for many centuries, during which the forests grew and died, forming a thick deposit of decaying, peat-like matter. At long intervals more abrupt land subsidence occurred, the forests and decaying matter were overwhelmed by water from sea or land, and the rivers once again poured layer upon layer of silt over the area. With occasional long pauses between one occurrence of subsidence and another, the water would become shallow again, new forests grew and died, and new layers of peat-like material formed. As the number of coal seams indicate, this cycle recurred frequently. Millions of years later these conditions came to an end. In time, the area of land and sea changed and, although a thousand or more metres of rock were formed on top of the rocks laid down in Carboniferous times, in Britain they never again contained extensive or numerous peat layers.

Right: A researcher at the N.C.B. Coal Research Establishment, Stoke Orchard, investigates the petrology of coal with an automatic reflectance microscope.

Peat to anthracite

The process of converting decaying vegetable remains to coal begins with the activities of aerobic (oxygen-demanding) micro-organisms in the sludge of the nearly stagnant waters of the swamp, leading to the eventual formation of a peat-like layer. Burial of this layer under a thick covering of silt stops not only plant growth, but also eventually bacterial action, although in the initial stages of burial the peat-like material would also experience further bacterial attack by anaerobic (non-oxygen-demanding) organisms. Bacterial action would cease completely when the accumulation of layers of silt is sufficient to prevent the removal of decay products which are toxic to bacteria. The gradually increasing weight of the accumulation of inorganic material would squeeze water from the peat-like layer, so consolidating the peat. The pressure-effect of the weight of overburden, coupled with other pressure effects resulting during the course of geological time from massive movements of the earth's crust and including the rise in strata temperature due to depth, induced profound changes in the buried peat, giving rise to several types of coal, the peat-to-anthracite series, each sequence of which represents the extent of change or metamorphism of the original peat deposit. The different types of coal in the series are arranged as follows in the order of their increasing metamorphism from the original peat: peat → brown coal → sub-bituminous coal → bituminous coal → semi-bituminous coal → anthracite; each member representing a greater degree of maturity than the preceding one or, in short, a higher rank of coal. Anthracite is therefore the most mature and highest rank coal.

Outline of coalmining

Although there is evidence to show that coal was mined on a small scale in Roman times in areas of Britain where coal seams outcropped (i.e. are exposed on the surface), there is nothing to indicate that the Romans were interested in coalmining. With the collapse of the Roman Empire nothing is heard of coalmining in Britain for many centuries. There are references to metal-mining in the Domesday Book (1085) but none to coal. Wood was the common fuel. Coal-getting probably began in this country in the twelfth century in areas where coal seams outcropped along the coasts, with the collection of loose fragments of coal eroded by the tides (sea-coal) and washed ashore on the beaches. Additional supplies would be obtained from other seams that outcropped inland, especially in coalfields where a shortage of wood forced men to dig for coal along outcrops in hillsides or valleys. When as much coal had been obtained at one site as could be safely extracted, the miners, realising that the seam continued far beyond the end of their burrows, sank shallow pits or shafts to reach it further from the outcrop; then they again burrowed into it to extract as much coal as they safely could. The coal was carried to the surface up ladders in baskets borne on the backs of bearers. From their finished shape, such excavations were known as bell-pits.

The shallow bell-pit, like the surface outcrop working, was more

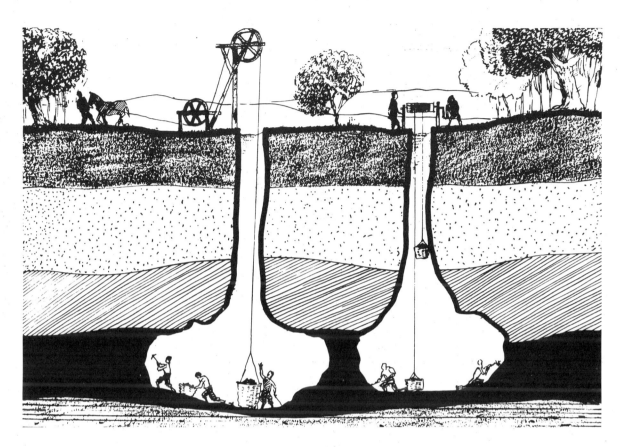

A cross-section of a bell-pit mine, the first system of underground mining.

akin to quarrying or opencast mining than to systematic deep coalmining which could be said to begin only when the more venturesome of the miners began to drive much further into the seam a series of inter-connected roadways to form coal pillars. These, since the workings at that period were usually at shallow depths, were left in place to support the overlying strata and avoid damage to the land surface. Mining in this fashion continued until the natural ventilation of the workings or the transport of coal became too difficult. The working was then abandoned, a new series of tunnels was driven into the seam from another outcrop site and mining as described above continued. Wet strata mainly confined the early miner to workings advancing to the rise so that water drained away naturally; such workings also made it easier for the drawers or bearers to drag the extracted coal on trays, or to carry it in baskets from the coal face to the mine entrance. Coal could be mined to the dip, provided a suitably placed adit, or drainage tunnel, had first been driven from the surface. This general method of working, in which neither machinery nor vertical shafts were required, represents the first stage in the evolution of deep coalmining.

By the middle of the thirteenth century coalmining in Britain had become a well-established industry which expanded slowly during the fourteenth, fifteenth and sixteenth centuries. More vertical shafts were sunk in which the coal was raised by hand-operated winches. Barrows

Left: An engraving, dated 1556, of three methods of ventilating mine tunnels and drawing off 'heavy vapours'. There are two sets of bellows (at A and N) and a third system driven by the horse on the treadmill (H).

began to replace trays and baskets for transporting coal underground. When it became necessary to exploit deeper seams to meet the growing demand, deeper shafts were sunk and the horse-gin, with its associated rope and drum, became common for winding coal in shafts. Mining at

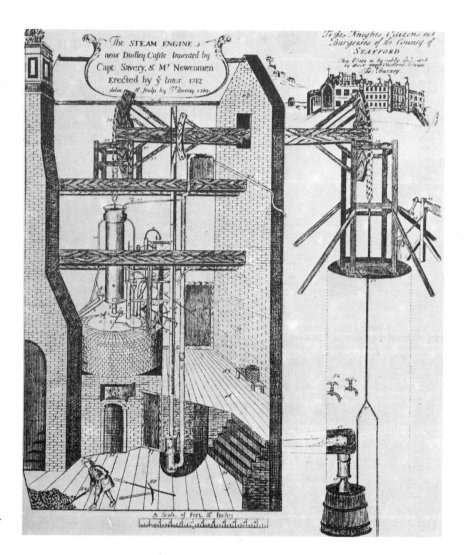

Right: A cutaway diagram of the first steam engine used in coalmining, installed at a coal pit in Staffordshire in 1712 for de-watering.

greater depths, however, brought more trouble from water and, by the end of the century, simple machinery, a chain-and-bucket system similar to the pot-chain device of the ancient Babylonians and Egyptians, was used for the removal of water from deep mines. Horses provided the power for operating the chain through suitable gearing. Coalmining had now reached another stage in its historical evolution when both coal and water were raised to the surface by machinery.

With the second half of the seventeenth century came a dramatic rise in the demand for and production of coal which continued through the eighteenth and nineteenth centuries. Annual output of coal, which was 2 million tons in 1660 and had reached 7 million by 1730, exceeded 200 million tons by the year 1900. The eighteenth and nineteenth centuries, covering the industrial revolution, saw many striking advances in the engineering technology applied to mining operations ancillary to coal-getting. They began with the introduction of the steam-driven pump for de-watering deep mines which, by the

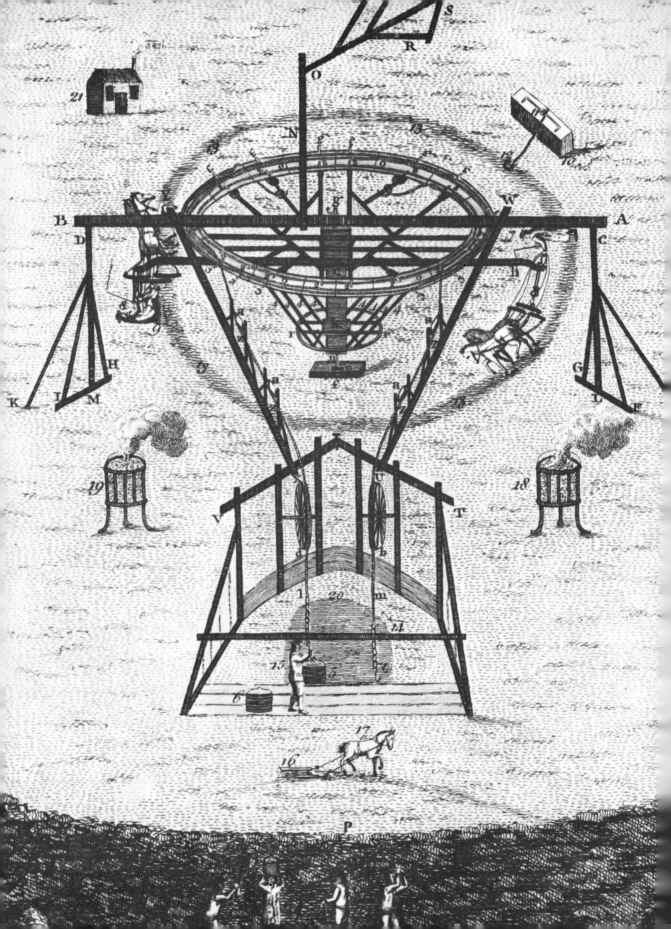

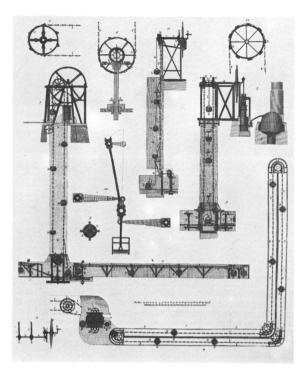
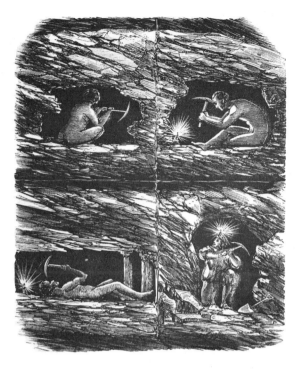

The steam-driven machinery above (dated 1801) for working coal mines was an improvement on the machinery seen below driven by a whimsey, an early nineteenth-century steam engine. This in turn had replaced the horse-driven system (left) in use until the previous century.

Above right: Until the late nineteenth century hewers worked in very cramped conditions. This illustration, published in 1860, shows both men and women at the face.

beginning of the eighteenth century had become the most pressing problem of the industry. There soon followed the application of steam power to machinery used for shaft winding and (later) for the transport of coal and mine ventilation, and for the compression of air and the generation of electricity, both of which were used for operating coal-cutting machines. Rock drills were also operated by compressed air.

By the end of the nineteenth century, nearly all of the coal was won by hand-pick, aided by explosives for blasting, and loaded by hand-shovel into trams or tubs at the working face. The methods of working the coal seams generally conformed with one of two basic systems, 'room and pillar' and 'longwall', usually applied within large rectangular panels pre-formed in the seam by driving suitably disposed development headings. 'Room and pillar', mainly adopted in seams exceeding 1.4 metres in thickness and occurring at depths of less than 450 metres, first consisted of splitting the seam into small rectangular pillars by driving two series of parallel roadways at regular intervals apart, and usually at right angles to each other, to form the pillars of the required size. Unless necessary for underground roadway protection or for the prevention of surface damage the pillars were later extracted (usually retreating from the boundary towards the pit bottom in a predetermined sequence) in a second operation in which successive lifts or slices were taken systematically from each pillar, the roof in the working place being timbered in accordance with established practice. As each lift was completed supports were withdrawn and the roof allowed to collapse.

Right: Miners going down at the start of their shift, in 1900.

Below: A miner in 1911 compacts a charge preparatory to blasting.

Left: Miners at the face hewing and loading into tubs (*c.* 1920).
Inset: Undercutting with a handpick.
Above: This miner, photographed in 1910, suffered severe deformation of the legs after years of hewing in low seams.

Left: A coal wagon of *c.* 1700 carries the coal from the mine head on the surface.

In 'longwall', applied in seams mostly at depths exceeding 100 metres and by far the most commonly adopted system in Britain, the coal is removed in one operation from faces varying in length from 100 to 250 metres. Until the application of mechanical face conveyors in the early years of the twentieth century, the longwall face was approached by branch or gate roads, about 12 metres apart, through which each miner and his helper reached their part of the face. These roads advanced with the face and support was given to the overlying strata by stone packs, 2 to 3.5 metres wide, built along the sides of each road from rock obtained by ripping stone from the roof in each branch roadway to enlarge its height. Additional roof support was given by setting props and bars in roadways. The coal was won from the face by the hard and tiring work of the miner, who usually undercut the coal by hand-pick before drilling shotholes for blasting the coal down with explosives. The loose coal was filled by hand into tubs or trams which were drawn by ponies along the branch roads to a main haulage road for transport to the shaft bottom by mechanical rope haulage. The roof at the coal face was supported on wood props and bars set by the colliers. As the working face and the roadside packs advanced, the back rows of face supports were withdrawn and the roof was allowed to collapse in the waste, or 'goaf', between the packs.

Mechanical progress underground

Coalmining is sharply distinguished from manufacturing industry by the high labour and labour-related cost content (approximately 55 per cent) in the prices it must charge for its products, which suggests that there must be ample scope for mechanical progress. Yet it must be remembered that the unique hazards of the environment in which coalmining operations are conducted impose many restraints on field

Above: A young boy in charge of a line of tubs pulled by a pit pony, in 1911.

experiments and trials to test new techniques and machines. The record shows that during the nineteenth century progress was slow; that during the first half of the twentieth century, despite the constraints imposed by two World Wars, progress was significant; and that, since nationalisation in 1947, progress has been dramatic, especially in coal face mechanisation.

(a) Underground transport

Following the earlier introduction of trams running on rails, progress in the mechanisation of underground transport started around 1800 when a steam-driven engine was installed near the pit bottom of a Tyneside colliery to operate a rope haulage which replaced several pit ponies and the men and boys in charge of them. Other installations followed, but high steam transmission losses and condensation troubles led to a search for a more suitable source of power for use underground. Eventually, in 1850, a compressed-air driven engine was installed underground to operate a rope haulage at Govan Colliery, near Glasgow. The use of compressed air, a safe power medium, soon extended and, although not nearly so handy and flexible as electricity which was introduced underground in the 1880s, it is still used today for certain purposes underground.

Following the introduction of power-operated haulage engines, there was little change in underground transport mechanism until the early 1900s, when scraper-chain, belt and shaker or jigger conveyors were developed for the transport of coal along longwall faces and in main gates. Some thirty years later more changes in coal transport at or near the face followed, with the introduction from the U.S.A. of duckbill and gathering-arm loaders (usually incorporating scraper chains) for loading prepared coal from room and pillar workings or

Left: Compressed air was first introduced underground in 1850 to power a rope haulage system. Here is shown a more modern use of compressed air, an early experiment of pneumatic transport of coal in pipes at the N.C.B. Coal Research Establishment, Stoke Orchard. The bends in the pipes are deliberately smooth to prevent blockage.

from the face of exploratory or development headings, and delivering it to outbye main conveyor systems or, occasionally, into large modern mine cars, having a capacity of up to four tons and fitted with brakes, springs and roller bearings and designed for automatic tippling without one car being uncoupled from another.

The most important single contribution to the efficiency of underground coal transport came from the engineering development of the belt conveyor. Indeed, the capacity of trunk belt conveyors now makes a drift from the surface more satisfactory for raising coal than a vertical shaft, even when mining at depths from 500 to 750 metres. Conveyor belt installations in surface drifts exceeding a mile in length, however, are not satisfactory for manriding and so the best compromise is a combination of surface drift for coal transport and vertical shafts for men, materials and ventilation. In the late 1940s came the

Right: A 1910 coal tub rotator, the predecessor of the modern automatic tippler for mine cars.

development of the cable-belt conveyor for long hauls up relatively steep roadways, in which the belt, used merely as the carrying medium, is supported and carried along by wire ropes.

Efficiency in the operation of long trunk conveyor installations has been improved by the use of bunkers for the storage of coal and for regulating the flow of coal to the belts, allowing them to run fully-laden for as long as possible. In some trunk conveyor systems, additional manpower economies are obtained by using long heavy-duty conveyors to eliminate transfer points and by the introduction of remote control. Transfer points are designed to reduce degradation of coal and the risk of blockage, while monitoring equipment and closed-circuit television are provided to guard against excessive temperatures, blocked chutes and torn belts.

Locomotives, which operate on gradients of less than 1 in 25 are not extensively used underground for the transport of coal. Their use for transporting men and materials, however, is extending. Diesels are used for the steeper gradients and battery locomotives, extremely reliable and relatively easy to maintain, are used for the flatter gradients. Considerations of cost and safety have not encouraged the use of electric trolley locomotives.

Conveyor transport has solved the problem of good clearance of coal from the face, but not that of manriding or the movement of materials. Detailed studies of these problems suggest that colliery layouts should be designed on the basis of rapid, convenient and continuous manriding from the surface in drift mines, or from the shaft bottom in other mines, to the gate-end of every longwall face, and that the earlier concept is changing which puts the emphasis on a main road for coal haulage and other roads for men and materials. Near to the

Above: Picking out the coal from the stone on a screen in 1901. The coal is fed in at the far end of the picture.

face the monorail system of transport has now become established; trackless vehicles and 'ski-lifts' for manriding are sometimes used.

(b) Mechanisation of coal getting

The first important steps to mechanise coalcutting were taken during the 1850s, when attempts were made at a few collieries to undercut the coal on longwall faces with a heavy coalcutting machine powered by compressed air. The first cutting tool was a revolving steel disc; its perimeter was studded with removable picks which were renewed as they became blunted. In later trials a pick-studded endless chain was used and, later still, a pick-studded rotary steel bar. From these prototypes evolved the disc, bar and chain coalcutters. Although at first disc machines were widely used as well as an appreciable number of bars, few, if any, of these types have survived. Chain-machines became the most popular and are still in use. Following the first trials, at least twenty years were to pass before the heavy coal-cutter achieved a reasonable measure of success. Much of the early experimentation was dogged by problems with the transmission of compressed air power to the machine. But the arrival of electricity in the 1880s stimulated mining machinery manufacturers to take a much greater interest in the design of coalcutters, and during the later years of the century their use gradually extended. By 1900 there were 180 at work.

During the years from 1900 to the outbreak of the First World War there was a growing demand for coal both at home and for export; in

1913 coal mines produced their highest annual output of 288 million tonnes. The use of coalcutters, helped by the development of higher-grade and special alloy steels and of more robust, compact and reliable coalcutters, increased. It soon became evident, however, that to achieve the potential of the machines it was essential that the machine-cut coal could be quickly and regularly removed. This brought the development of the mechanical face-conveyor, the first of which, the Blackett scraper-chain, was installed at a Durham colliery in 1903, to be followed a year or two later by a Sutcliffe belt-conveyor at a Yorkshire colliery. About this time, also, the shaker or jigger-conveyor, developed in Germany, was brought into use in Britain.

The introduction of the longwall face conveyor marked another major advance. It brought many advantages. By eliminating the need for branch roads, it reduced the amount of stonework required; it increased safety because it automatically called for a straight face, thus eliminating the 'stepped face' with its corners and heavily stressed roof zones; and it eased the introduction of a systematic roof support system. Between the two World Wars the combination of the coalcutter and the face conveyor became standard practice and was usually operated on a daily three-shift cycle system, known as 'conventional' longwall. On one shift the cut coal (prepared by blasting where required) was loaded by hand on to the conveyor, and the preparatory work for the next day's loading was conveniently spread over the next two shifts. Between 1900 and 1950 the proportion of machine-cut coal rose from 1.5 per cent to 79 per cent; and for coal face-conveyed from 0 per cent to 85 per cent.

Then in the 1930s came the Meco-Moore cutter-loader, essentially a coalcutting machine trailing a loading unit behind it. The machine, travelling in one direction along the face on the floor in front of a conveyor, undercut the seam; the coal was blasted down; and, as the machine travelled back along the face, the prepared coal was loaded on to the face conveyor. A promising start had been made on the development of a machine for cutting and loading coal simultaneously.

The Second World War created a huge demand for coal at a time of critical manpower shortage, and therefore a demand for machines that could load coal. Several American machines designed for narrow work were introduced: the Duckbill and Joy loaders, and a heading coal-cutting machine which loaded coal either on to a conveyor which advanced with the machine or into a shuttlecar, a trackless type of battery-powered vehicle low-mounted on rubber tyres, which shuttled backwards and forwards between the face and the point where the coal was transferred to the main outbye transport system. British manufacturers developed other longwall loaders, such as the Shelton and the Huwood. But these machines, designed only to load cut and prepared coal, fell short of the ideal, a machine that cut and loaded coal simultaneously.

The pressing wartime demand for more and more coal brought a reconsideration of the original Meco-Moore which resulted, in 1943, in the production of the wide-web AB-Meco-Moore, the first longwall machine to cut and load coal simultaneously. It achieved a remarkable success. Other British cutter-loaders followed: the Logan Slab-Cutter, the Uskside Miner, the Gloster-Getter. Across the Atlantic, the Americans produced continuous miners – the Lee-Norse and the Joy-Miner, and the Canadians the Dosco, for longwall faces and for cutting and loading coal in room and pillar workings or in driving exploratory, or development headings in coal. From Germany came the coal plough and, more important, the armoured flexible conveyor (A.F.C.), on the face side of which the plough was mounted (see Chapter 3). Both were introduced in Britain after the Second World War. Hauled across the face and back again by its wire rope hauler, the plough wedged off thin slices of coal on to the A.F.C. Two British cutter-loaders based on this wedge principle were also developed about this time, the M. & C. Samson-Stripper and the Huwood Slicer-Loader.

Despite their success, the use of wide-web longwall cutter-loaders like the AB-Meco-Moore and Dosco tended to weaken the roof – by leaving a wide span unsupported – and increase the risk of accidents from falls or of excessive dust – which could allow coal dust explosions or add to the potential hazard of pneumoconiosis.

Progress since Nationalisation
(a) The prop-free-front face
Help in meeting many of these disadvantages was, however, soon forthcoming. The decade that followed the Second World War, saw spectacular developments. One was the introduction of the previously mentioned German plough and its A.F.C., and another was the British invention of the narrow-web cutter-loader, the Anderton Shearer, and the AB-Trepanner. The hard nature of British coals limited the German plough's field, but it soon became evident that the A.F.C. could be strengthened to provide a track for the narrow-web coalcutters with many advantages (see Chapter 3), so making practicable that dream of mining men, the longwall prop-free-front face, along which the cutter could travel without obstruction and across which the flexible conveyor could be advanced without dismantling.

(b) Self-advancing powered supports – total caving
This modern system of mining has contributed to efficient and safer mining in other ways. Because of the higher rate at which the power loader advanced along the face, it created a need for more flexible types of roof support and concentrated study of the development of strong types of support which allowed ready and reliable means of variation in length, rapid setting, pre-loading, speedy release and ease of advancement; all were gradually accomplished. Initially, the new system led to a wide use of friction-type yielding props and, in the late 1940s, to the introduction of the more successful hydraulic prop. Success in the use of this prop and further developments in its design led, by the mid-1950s, to the development of self-advancing systems of powered-roof supports. At a time when mining tech-

Above: A model of the Anderton Shearer Loader of the late 1940s.
Right: The Anderton Double Jib Shearer, Mr Anderton's first try.

Right: In this cross-section of the face before the 1950s, the cutter cut horizontally into the coal: holes were then drilled and the explosives fired. The next shift shovelled the coal on to the conveyor, and the third shift moved the conveyor forward to the advanced face.

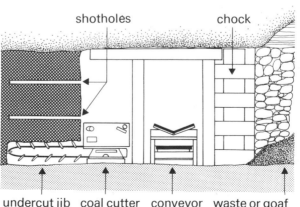

niques were rapidly changing, this new support system, because of its inherent strength and flexibility, proved to be of great value. By providing a strong 'breaking-off' force at the waste edge, it contributed to early collapse of the roof in the waste and led to a rapid extension of total caving of the waste; mechanising (by means of hydraulics) the movement of the whole support system eliminated much physical effort; and ensuring systematic support of adequate strength made for safer roof conditions and provided suitable conditions for mains-lighting the face. It also led to a dramatic reduction in accidents from falls of ground at the working face, to longwall faces producing 3,000 tonnes a day, and made the million-tonne-a-year face possible. It was these developments which, by greatly reducing the manpower required, brought about a real break-through in productivity.

(c) Remotely operated longwall face (ROLF)

Having in mind these dramatic advances, little wonder thoughts turned to the possibility of the complete integration and remote control of coal face operations with the elimination of manpower except for the purposes of installation, maintenance and inspection. Experiments on two faces to test the practicability of the remotely operated longwall face (ROLF) were carried out about ten years ago in two collieries in the East Midlands. Although not wholly successful, the results were sufficiently encouraging to warrant continued research. It would, however, be rash to promise the general adoption of ROLF at this stage, although such faces may be feasible in relatively thick seams in an area known to be free of geological disturbances or abnormalities.

(d) Geological exploration for reserves

The general aim of all these technological advances is to increase the rate of output. However, simply to maintain the total output at around 120 million tonnes per annum until the end of the century, a massive geological exploration programme had to be put in hand. The National Coal Board realised that there was not enough evidence on the availability of economically workable reserves to permit development plans to be made for the industry to meet its expected contribution to the national energy requirement. So during the past five years exploration has been stepped up with core-drilling and geophysical (seismic reflection) surveys. (Seismic reflection surveying consists of firing small shots in shallow drill-holes spaced at intervals along traverse lines, and recording the reflected vibrations so produced on sets of geophones placed in the same traverse. The recording equipment is advanced along the traverse to keep a constant relationship with each successive shot point.)

A major problem was the selection and timing of areas for exploration, a problem not eased by the magnitude of the exploration programme, coupled with a shortage of exploration equipment. Since the output from new collieries would, of course, take longer to provide than that derived from either the expansion of output or an increase in

The increased use of coalface lighting in the last decade has helped efficiency and safety.

the life of existing collieries, there was a tendency to go for the latter schemes. There is a danger, however, that if carried too far this policy might seriously damage the long-term needs of the coal industry, especially in the 1980s and 1990s, a period when world oil production may well be declining and when nuclear power will probably still not be in a position to provide a major source of energy demand.

(e) Geological exploration – centres of activity

The programme of exploration has so far provided much valuable information on economically workable reserves, ones that are either accessible to working collieries or in areas requiring new sinkings. The National Coal Board's activities have been concentrated in these areas: in Scotland, off the north and south shores of the Firth of Forth; in Northumberland, from Druridge Bay north of existing undersea workings; in several areas in Yorkshire, and especially in the district between Selby and York (see Chapters 4 and 5); in the West Midlands, on the fringes or between deep inaccessible Coal Measure basins; in the East Midlands, between Gainsborough and Newark, and especially in the Vale of Belvoir, beneath which lie at least 350 million tons of reserves.

In South Wales, the degree of exploration is limited to confirmation of the quality of seams and the intensity of structural deformation in confined areas. Seismic surveys have also recently supplemented preliminary boreholes in a new venture in Oxfordshire, between Oxford, Banbury and Stow-on-the-Wold, where several low- or doubtful-quality coals, high in the Coal Measures, lie below a cover of 600 to 760 metres in mesozoic rocks.

Since the average age of active coal mines in this country is about seventy-five years, exploration for additional reserves must continue alongside improvements in coalmining techniques if the size of the industry is to be maintained at an adequate level. Fortunately, the cost of exploration, when compared with that of developing a new mine, is relatively small. Its aim is not only to provide new coal resources of up to fifty years' output in reserve but also to discover resources of such quality, size and distribution as will justify the expenditure of large capital sums on their exploitation.

3. The New Mechanisation

E.L.J. Potts, Dr. R.K. Dunham and John Scott

The slow, first mechanisation of the mines, as we have seen in the previous chapter, was followed from about 1955 onwards by a more rapid movement, still in progress, that was forced on the National Coal Board by the low cost of imported oil at the time.

In only ten years coal had gone full circle, from the highly advantageous position of scarcity and practically zero stocks, to embarrassing surpluses with some 30 million tons of coal in stock. Something had to be done quickly to restore the competitiveness of coal. Rapid mechanisation, it was decided, would reduce production costs, to be followed eventually by closure of high-cost pits not worth mechanising.

Modern faces operated by cutter-loaders can get coal on all of their three daily shifts. The most profitable faces have a week of ten to fifteen shifts, a great advance on the six shifts of the old mechanised faces before the five-day week. Cutter-loaders have enormously helped in the increase of the British miner's productivity. It has risen from an average of below 1.1 tonne per manshift in 1947 when the mines were nationalised, to 2.29 tonnes per manshift in 1975.

These productivity figures are for all the men in the colliery but those for productivity at the coalface itself are even more favourable to the modern miner. His output averages nearly three times what it was in 1947. Since the modern coalface can also get coal on fifteen shifts instead of six, its output per week can be many times more than it was then. But this high productivity has to be paid for and the investment cost for one modern face can reach £500,000. In 1947 the 'modern' face was equipped with only a coalcutter and a conveyor, probably costing not more than £20,000.

Before describing in detail the various face machines in use now, we should give a general explanation of four features of the modern coalface that make it different from and more productive than the less mechanised face of the years up to 1950, with its coalcutter and conveyor. These features are the prop-free-front principle of roof support, a radical departure from previous safety practice, that required the permission of the Mines Inspectorate before it could even be tried; secondly, the armoured face conveyor (A.F.C.); thirdly, powered roof supports; and fourthly the cutter-loader itself, the all-important face machine that passes up and down the face, removing coal like a bacon slicer.

Right: A simplified plan of a modern mine, showing three advancing longwall faces and the development roads of a fourth face. To ensure an efficient airflow the intake airway must be separated from the return, by means of air crossings. The main intake airway is usually the main haulage route for coal and contains a trunk conveyor.

Above: An example of the clean look of the typical modern mine – Parkside colliery in Lancashire.

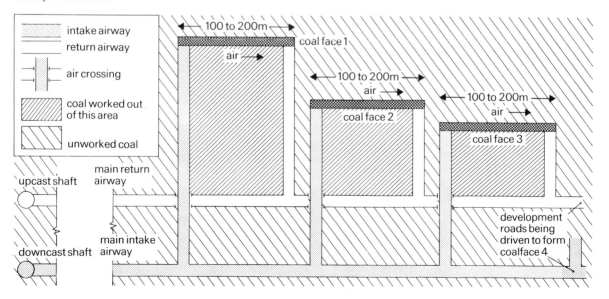

The prop-free front

In early mechanised faces, the front row of props had to be set within 3ft 6in. (about 1 metre) of the coalface because the men worked close up to it. The coalcutter passed up (or down) the face in a space 2ft 6in. wide, in front of this forward row of props with only the protection of the overhanging wood or steel bars. The idea of a prop-free front, with up to 6ft 6in. (about 2 metres) of roof unsupported in front of the row, had to be approved by the Mines Inspectorate for safety reasons.

It was illegal, and at first sight seemed dangerous, to allow so much unsupported roof between the coalface and the nearest row of props. But this unsupported area is occupied only by a machine, not by men. The machine travels fast, and consequently the roof is unsupported for only a short time. The powered supports are moved in quickly by hydraulic power without men having to pass under unsupported roof. All today's coalface machines work on a prop-free front and it is now almost impossible to imagine a time when this was not so, since the face conveyor has to be in close contact with the coalface.

The modern continuous mining methods introduced with the prop-free front have greatly helped mining engineers in their efforts to concentrate mine workings. All mines have the disadvantage compared with a factory that their underground workings are spread out and consequently difficult to supervise and expensive to run. Modern power-loaded faces, on the other hand, provide higher outputs from a similar area. With them, spare face room can be made by double-shifting a face that was formerly worked on one shift or even by treble-

Below: How powered supports work. The upper unit is so designed that, by operating control valves, the A.F.C. and support units are moved forward to their new positions without hindrance to coal getting.

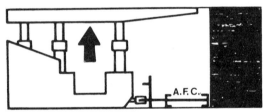

STEP A
The vertical supports are extended to bring the roof beam into contact with the newly exposed roof.

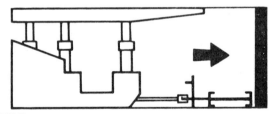

STEP B
The power-loader has passed on the loading run and the double-acting ram is extended, pushing forward the conveyor.

STEP C
The vertical supports are lowered.

STEP D
The ram is set into reverse and the support is drawn up to the new face line. It is then reset to the roof (step A).

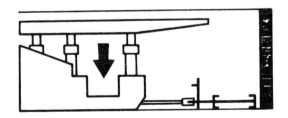

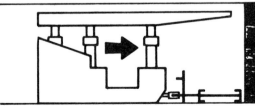

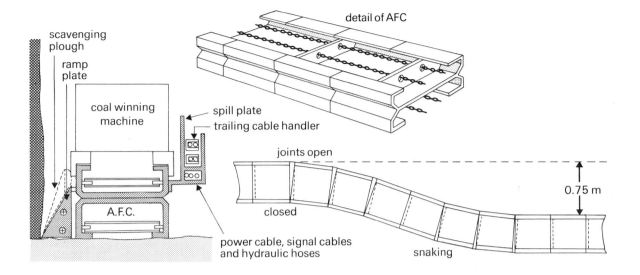

Above: The armoured face conveyor is made up of sections loosely linked together so that they can be snaked sideways or in a curve. It carries the cutter-loader, and its endless chain or chains with flights attached move the coal back to the main conveyor.

Overleaf: Cutaway layout of the modern coal mine with faces advancing away from the shafts.

shifting a double-shifted face. It must be admitted, however, that a face regularly working three shifts a day, fifteen shifts a week – almost continuous coal-winning – is not always a happy one and managers may prefer to get coal on two shifts, leaving the third for essential maintenance or repairs.

The armoured face conveyor (A.F.C.)

Another essential of the modern coalface is the armoured face conveyor (A.F.C.), sometimes called the 'snaking conveyor' because in plan it can be curved round (snaked) behind the cutter-loader as it advances along the face. The A.F.C. is an all-steel trough structure running the length of the coalface and strong enough to carry the several-tons weight of the cutter-loader that slides along the top of its trough. An endless chain or two chains within the trough pull steel paddles (flights) that drag the coal to the end of the trough, where it is discharged to the roadway transport system. The flow of coal is there turned at right angles as it meets another steel chain conveyor, the stage loader. This in turn delivers on to a belt conveyor, and the coal is then well on its way to the mine shaft.

The A.F.C. has the advantage that it suits the cutter-loader admirably and can easily carry the heavy load from it. The coal can pass under the traversing machine in the trough of the A.F.C. and this can be promptly 'snaked' close behind the cutter-loader, enabling the props to be advanced only a few minutes after the coal has been cleared by the face machine.

Powered supports

Almost all movable roof supports underground are now hydraulic. They are set or released by merely opening or closing a valve that controls the flow of a hydraulic pressure fluid. The principle of the

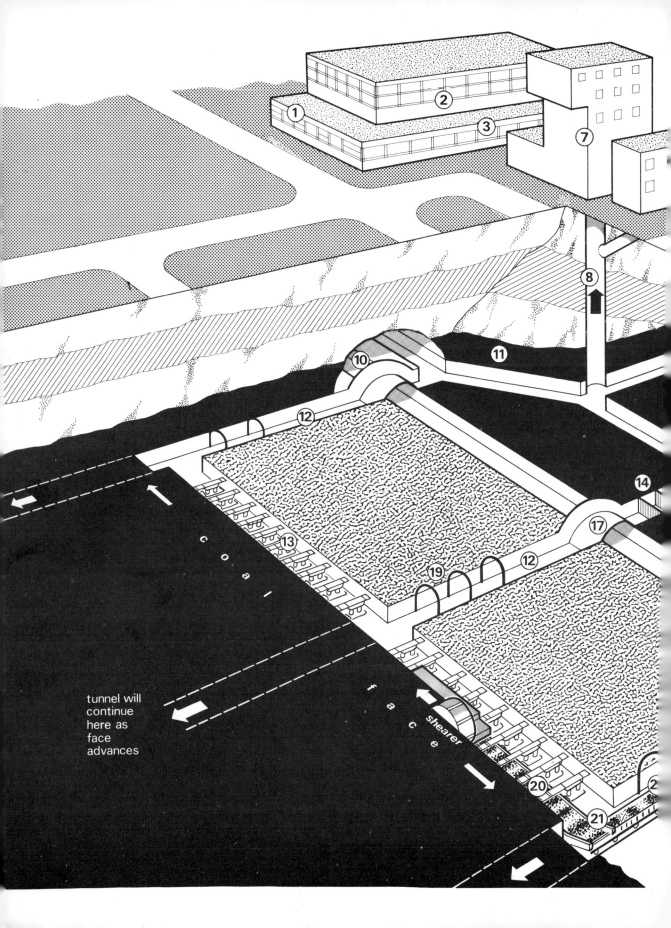

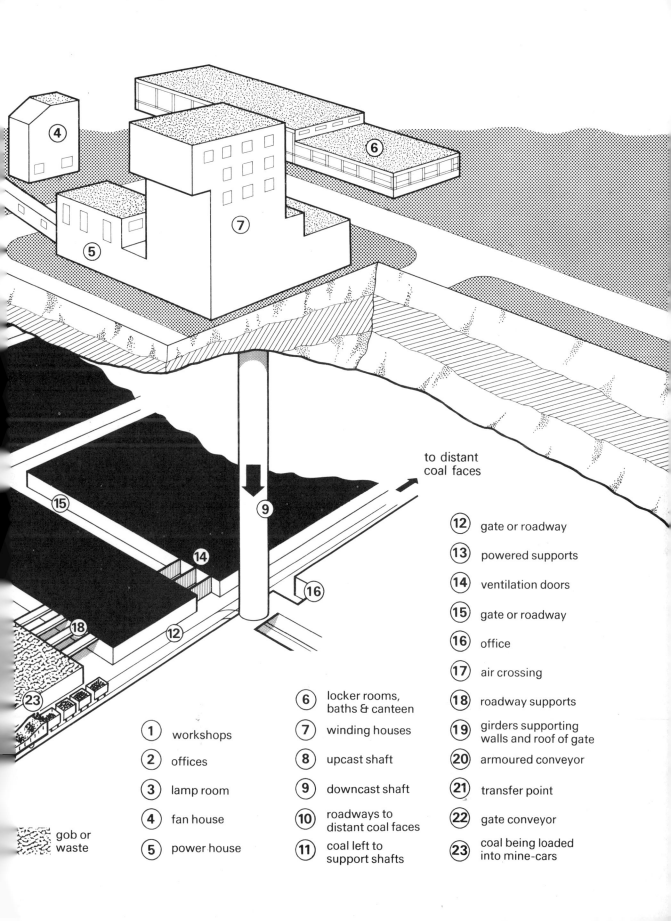

Left: Powered supports in a thick seam are moved forward by a miner. The face is to the left of the picture.

Below: Six-leg supports stacked, ready to go underground.

hydraulic prop is that its two telescoping sections are forced between the roof and the floor by the action of a pumped fluid (the 'hydraulic' fluid) on a ram. The fluid causes the ram (connected to one 'telescope') to move along a cylinder (in the other 'telescope'). The hydraulic fluid is non-flammable, usually 95 per cent water, 5 per cent oil. When, as is usual on the coalface, several of these props (occasionally up to seven) are grouped together and joined at the top to a roof unit and at the bottom to a floor plate, they are provided also with a horizontal jack linking the floor plate to the A.F.C. This enables the group of props to be advanced by hydraulic power and also enables the A.F.C. to be pushed forward. Such 'powered' or 'self-advancing' supports are almost universal in modern mechanised coalfaces.

The face equipment is advanced during the cutting run of the coalface machine by the four operations that are usual with powered supports:
1. pushing the conveyor forward by extending the horizontal jack
2. releasing the pressure on the support unit, lowering the props
3. advancing the support unit by pulling on the conveyor
4. applying pressure on the props again, tightening them to the roof

Since, for one group of supports, all these operations are completed in a few seconds, there is only a very short time when the roof is unsupported, and during this time the units each side of the released one can carry its share of the load. Years of use of these props since 1958 have demonstrated that safety at the face has enormously improved.

Cutter-loaders
The chief machine in the coalface is the cutter-loader. This powerful machine travels along the coalface in contact with the coal, usually sliding along the top of the A.F.C., breaking out a strip of coal about 560mm (22in.) wide all along the face and pushing the broken coal on to the conveyor. The machine is hauled by one of many methods – originally a steel rope, superseded by a chain, running the full length of the face and anchored at each end. The chain in 1976 began in turn to be superseded by chainless haulage methods that avoid the danger of injury from chains breaking or whipping under tension. Chainless haulage is often based on a rack and pinion mechanism, between the cutter-loader and the A.F.C.

Modern face machines, designed to work on a prop-free front, make narrow cuts of 550mm (22in) at the most so that the area of roof exposed at any time is not large. Before the prop-free front was introduced, undercuts of 1.5m (5ft) or more into the coal were often made by coalcutters.

For a cutter-loader to begin work in the face there must be a gap, or rebate, where coal has been removed, at the end of the face between the A.F.C. and the coal, called the stable. Originally stables had to be cut out by hand but this work is being progressively mechanised or indeed eliminated. Stable elimination is important because without it the aim of continuous mining can be achieved only by the expenditure of heavy manual labour in the stable.

Above: An M.R.D.E. chainless haulage applied to an Anderton shearer on display at the surface.

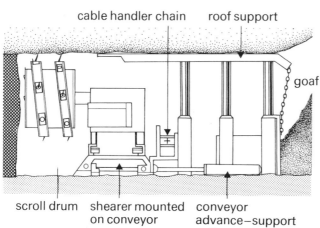

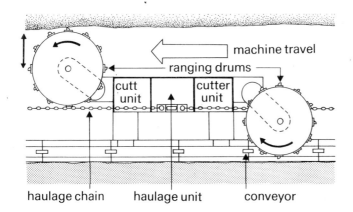

Left: Two views of the Anderton shearer, along the face (above) and from the face side (left).

All cutter-loaders exist in many different models for different conditions. Electrical power is supplied to them by a trailing cable connected to a distribution point, the 'gate-end box' at the intake end (gate end) of the face, at 550 to 1,100 volts.

The main gate or intake airway
There are two permanent access tunnels to every production coalface, one at each end. The intake tunnel (gate or roadway) is at the end of the face by which the fresh air enters, while the return (return airway) gives access to the other end, by which the air leaves the face. The main access is at the intake end and contains the electrical supply switchgear and the hydraulic tank, pump and motor that operate the powered supports and advance the A.F.C. A considerable amount of space is taken up by this electrical equipment, so the intake must be driven to a fairly large cross-section. The coal has to leave the face through it and it also must be travelled by men.

Electrical equipment is placed in the intake airway because the methane content there is naturally lower than in the return airway. Regulations forbid electrical equipment to be used when the methane content of the general body of the air exceeds 1.25 per cent. When the electric motors are stopped virtually all work on the face ceases, so no coal comes out.

Delivery of the coal from the A.F.C. in the face on to the stage loader is also at the intake end of the face. This short steel conveyor can extend forward easily, allowing delivery of the coal on to the longer, main conveyor system out of the mine. Also near the delivery end of the A.F.C. is its drive motor. This concentration of equipment where the intake access road meets the face, creates congestion there. Partly for this reason it is often difficult to keep the roadways at each end advancing as fast as the face. Yet if they do not do so, they delay the face work, reduce output and greatly increase production costs.

Coalfaces are very much more productive than they were in 1950, consequently they advance more quickly. Exceptional faces advance an average 3m (10ft) or even more per shift, and it is rare for a satisfactorily productive face to average less than 1m (3ft) per shift. Usually a small face advance is caused by the seam being thick, say 2m (6ft 6in.) or because it is just starting up. One exceptional face at Silverdale Colliery in 1974, however, in a thick seam of 2.4m (8ft) regularly advanced 2.6m (8ft 5in.) per shift over ten shifts a week. This condition of thick coal and high face advance resulted in an excellent face productivity of over 75 tonnes per face worker and a face output of over 2,500 tonnes a day or 1,300 tonnes per shift – which is as much as the output of some entire mines! This brings out one point about mining, the almost incredible differences in conditions and performance between one mine and another.

It has always been difficult, with mechanised mining, to ensure that the access roads advance at the same rate as the face they serve, and this difficulty has been accentuated by modern rapid face advances. Mining does not in this way differ very much from what it always was,

Left: An in-seam miner developed at M.R.D.E. seen at the surface. The machine advances into the coal, scoops up the cut coal and delivers it to the conveyor on the left. Stability is provided by two hydraulic props in the centre, which form the base from which the machine advances.

an industry where there is not enough space to do what is needed. This is brought home to the mining engineer when he tries to increase the rate of advance of his tunnels. Consequently some roadways are now being driven by tunnelling machines like huge horizontal borers.

Retreat mining

The difficulties of maintaining the rate of advance of the access roads to the face have caused many modern mines to drive them completely beforehand. This method, known as 'retreat mining', has been adopted on a large scale in Poland and the U.S.S.R. Where it has been possible to adopt it in Britain it has usually resulted in high productivity and consequently profitable mining. But it involves high investment costs and a much longer time for coal to begin to be produced from any one district, so the financing charges are high. If the finance is not available, retreat mining cannot be used. Under private enterprise finance was expensive so retreat mining was almost unknown before 1947 in the United Kingdom.

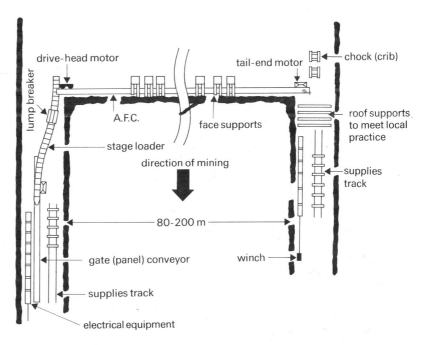

Below left: Setting up a new face with powered supports in the wake of an M.R.D.E. in-seam miner.
Right: A simplified plan of retreat mining, where two access roads are driven to the edge of the face (top of plan) and the coal between them is then extracted retreating back, the reverse process of an advancing face.

The reason for the delay is that the two access roads must be driven for the full length of the panel of coal that they have to extract, before the face starts to work. But these long drivages do provide almost complete information about the conditions in the coal that is to be worked. Indeed this is one of the greatest advantages of retreat mining.

Space at the face of a tunnel drivage is limited, so there are limits to the amount of equipment as well as to the number of men who can effectively work there. Even if teams of men are working at the face for 24 hours a day, drivages are still slow. But the tendency in the United Kingdom in 1976 was to encourage mine planners to use retreat mining where depths and consequent problems of overburden pressure were not excessive.

Ripping and packing

Retreat mining, however, is not typical at the moment of British coal mines and a common difficulty in keeping the roads up with the face is caused by the need to dispose of the stone that is broken in ripping.

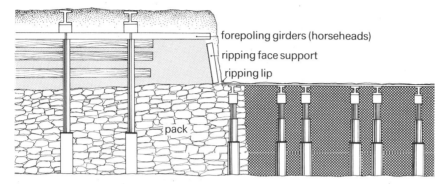

Left: Cross-section of a ripping face and lip. The coal has been extracted below the ripping lip, and the stone above it will then be broken out of the roof by ripping to make more height. The girders hold supporting arches while they are being set in place, as the face moves forward.

Since the intake has to house bulky equipment, including a belt conveyor, and must be high enough for men to walk comfortably along it, stone has to be broken out of the roof or the floor so as to make more height. The usual method, breaking it out of the roof, is 'ripping'. Breaking it out of the floor is 'dinting'.

Where is this stone disposed of? In some exceptional collieries it has been possible to dispose of the stone separately by storing it in an underground bunker, and then at a suitable time stopping the coal flow on all belts, allowing only stone to travel out on them. This has been done where computer control of conveyors has been introduced. Where it is possible it is an excellent arrangement because the stone goes straight to the tip on the surface and does not overload the

Left: An operator controls an underground conveyor system by a mini-computer on the surface.

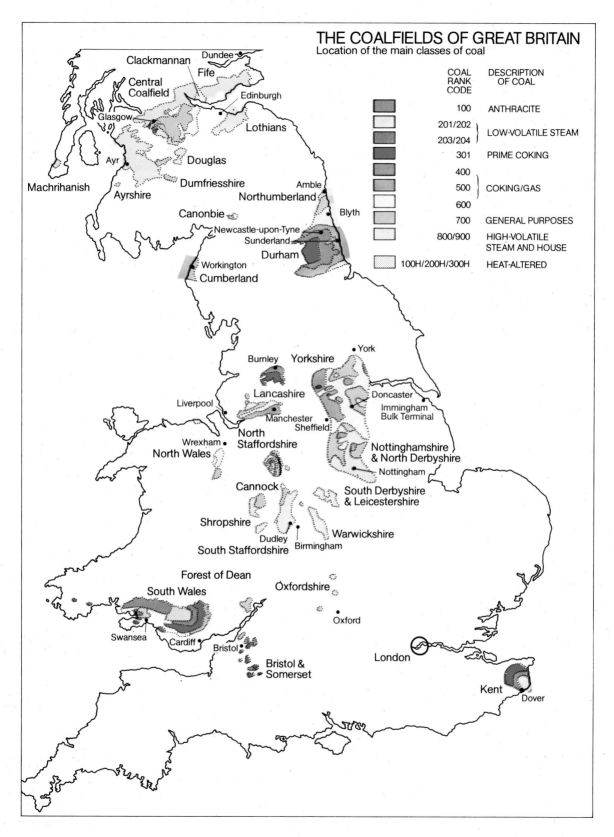

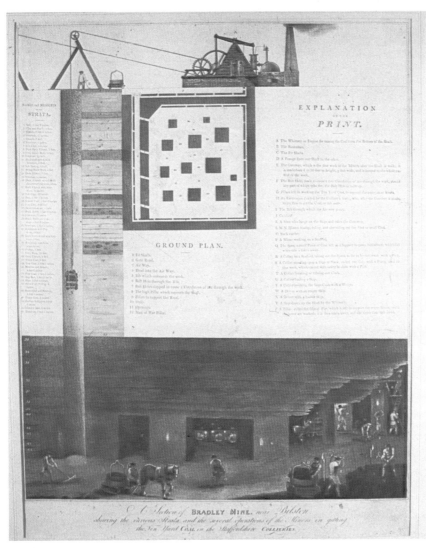

Above: An engraving of a cross-section of the ground plan of the Bradley mine in Staffordshire in the early nineteenth century.
Right: This sloping tunnel, or drift, at the Prince of Wales Colliery, near Pontefract, Yorks, is one of a pair being driven in the development of a mine first sunk in 1872. The life of the colliery will probably be extended to beyond AD 2000 at a productivity rate that should be more than four times the national average.

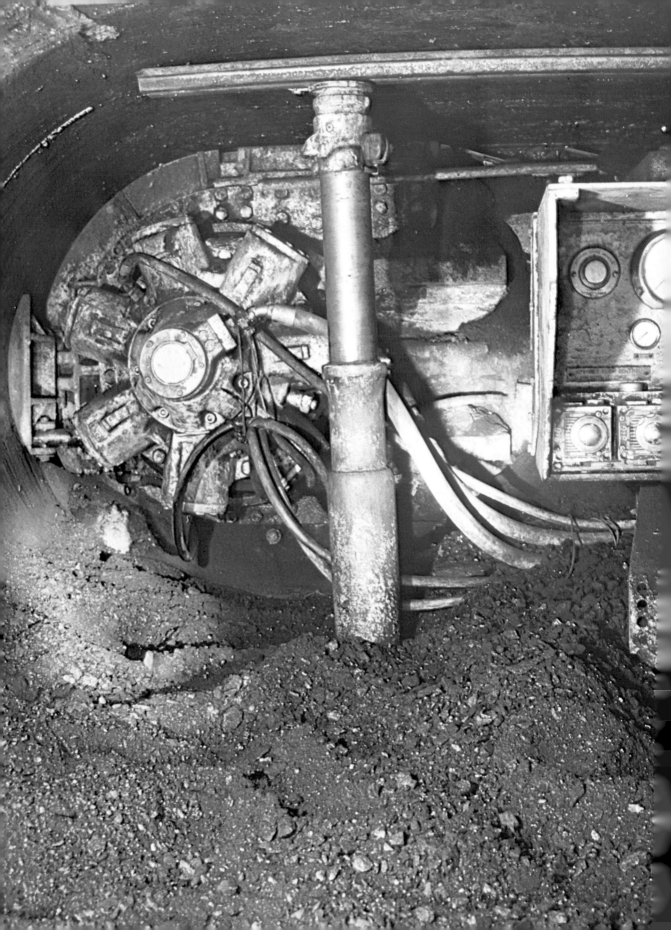

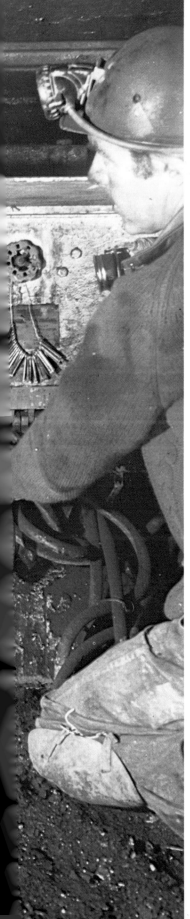

Above: A mine rescue team training underground.
Left: An M.R.D.E. in-seam miner in operation (see Chapter 3).
Below: Miners at Dawdon Colliery, County Durham, come up from their shift.

Bevercotes, near Worksop in the North Nottinghamshire coalfield, was first sunk between 1953 and 1958. After a period as a testbed for advanced mining, production began in July, 1971. 700,000 tons a year is presently mined from its Parkgate seam, and it is expected to produce in the future 1½ million tons a year. Left: The surface drift to the Parkgate seam. Right: A minicomputer for controlling surface operations.

Left: An NCB-Thyssen FLP 35 full-face tunnelling machine of 684 horsepower, which bores a tunnel of 3.65 metres diameter and weighs 98 tonnes.
Above: The base of a drilling rig in operation during the programme of exploration around Selby.

Left: A core brought up by a drilling rig in the Selby area.
Below: A surface tester establishes the location and depth of the coal seam.
Overleaf: A boiler at the Coal Research Establishment, Stoke Orchard, fitted with a fluidised shallow-bed combustor to show how a boiler can be converted to burn coal with an increased output.

The town of Selby, its abbey and the river Ouse that runs through it will be safeguarded by the decision to leave a thick pillar of coal beneath the town and rigorously to control the degree of subsidence.

Selby is a thriving industrial centre. Left and below: The mills of Ideal Flour and Hovis.

Right: At the ripping lip two new steel arches are set over forepoling girders, which carry their weight. The girders are temporary supports which advance with the ripping lip, and when the arches are in place they in turn carry the girders as they are moved forward to allow new steel arches to be put in place.

washery. In other collieries, the stone goes out on the belt mixed up with the coal and is removed in the washery. In most collieries the stone is disposed of entirely underground in the traditional way by building 'packs' or stowing mechanically.

Packs are thick stone walls built tight up to the roof alongside the edges of the access roads, which help to support them against the rock pressure. They may be as much as six metres thick. The difficulty occurs when the face of the tunnel is well ahead of the coalface (as it should be) and the broken stone has to be transported a long distance back, behind the A.F.C. Many methods exist for getting over these difficulties, some of them involving packing machines that convey the stone and press it tight against the roof. One system involves screening the broken stone and pumping into the pack the smalls that pass through the screen. A cement-water mix is blended with the stones just before they enter the pack, to make a crude but adequate concrete, held in by rough shuttering. The old method of building packs by hand meant moving the stone entirely by shovel, and this still has sometimes to be used.

Cutter-loaders
The Anderton Shearer Loader
About 75 per cent of coalfaces in Britain now use shearers, another 22 per cent use trepanners and 2 per cent use other methods, mainly ploughs.

The shearer was introduced in 1954 and was so quickly adopted that in 1958 it was producing 45 per cent of all mechanised output. It had only a 50 horsepower motor, compared with modern 400 horsepower versions that can cut and load 800 tons of coal per hour. The cutting head is a rotating drum equipped with picks that break the coal, and vanes that throw it on to the conveyor. The first shearers could cut in

Left: The type 500 Anderson Mavor ranging-drum shearer is an unusually powerful face machine.
Below left: A ranging-drum shearer in action.

only one direction and had to return to the end of the face before they could start cutting again. Bi-directional shearers were then developed that could cut in either direction. Another refinement is the ranging drum shearer, with a drum fitted on an arm that can be moved up or down to cut a thick seam – used in seams of more than 1.35m (4ft 6in). Yet a further development has been to put a shearing drum on each end of the machine, extending the capability still further in suitable face conditions.

The Trepanner
Shearers have the disadvantage that they produce small coal; this characteristic is difficult to overcome. The other main machine used in Britain is the Anderson Boyes trepanner. It makes larger coal but is more complicated than the shearer. It has several different cutting heads; the main one, producing the large coal, is cylinder-shaped so that it cuts a cylinder of coal in the same way as the surgical instrument of the same name designed for cutting into the human skull. The trepanning wheel is 0·85m diameter (34in). There is one at each end of the machine so that it can cut in both directions. The trepanner also has separate jibs for floor cutting, roof cutting and back shearing. Shearing is the operation of making a vertical cut in the back of the coal. Trepanners were originally floor-mounted and could consequently cut in thinner seams than was possible for shearers, but many trepanners are now, like the shearer, conveyor-mounted.

Below: The main cutting wheel (seen on the right) of a trepanner, throwing coal on to the A.F.C.

Machines for tunnelling, ripping, dinting or packing

The National Coal Board, through their Mining Research and Development Establishment at Bretby, have for years been spending money on research for investigating tunnelling machines and their development, as well as on machines for ripping, dinting and packing. The reason is that only too often, a productive, profitable coalface is delayed and output reduced because of difficulties in advancing one or other of the two access roads.

The worst trouble is likely to occur at the intake end of the face because of the vast quantity of equipment installed there, especially the electric motors and gear boxes of the A.F.C. and the stage loader, creating congestion just at the most awkward point, the delivery end of the A.F.C.

Retreat mining does not eliminate the congestion but is no longer such a problem because there is no need to build packs or remove many cubic metres of broken stone. Consequently there need be no delay to the advance of the face, though difficulties can occur. The access roads may be squeezed by roof pressure before the face reaches them and they may then have to be enlarged and re-made, which is always expensive.

Conventional advancing faces will probably always be used despite the advantages, in good conditions, of retreat working. N.C.B. research has therefore widely investigated all the face-end possibilities. The conventional way of driving the roads at each end of an advancing face was, and is, to drill holes above the coal after its removal, and to blast down (rip) the required amount of stone above the seam.

Ripping

Ripping machines are commoner than full-face tunnellers, which are much larger and more expensive to develop, build and install. Ripping machines replace blasting but generally can be used only in the weaker rocks.

The two types in use both have a hydraulic boom. One type has a rotating cutting head on the end of the boom; the other, the impact ripper or 'woodpecker' has a pecking action. Most machines are caterpillar-mounted and provided with a scoop in front so that they can push the scoop under a pile of broken rock and load it out.

Right: A woodpecker (or impact ripper) heading machine. This type hangs from roof girders.

Below right: A rotating-head ripping machine, known as a Dosco Roadheader, mounted on caterpillar tracks, has its own broad loading scoop and conveyor (foreground and right centre) to remove the rock it breaks. At the upper left is the canvas tubing through which fresh air is blown into the face of the roadway.

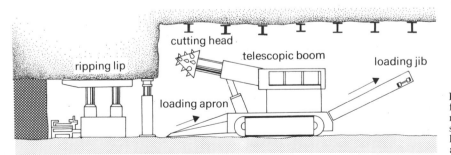

Left: Cross-section of a ripping face and a rotating-head ripping machine. The broken stone is scooped on to the loading apron and is carried away via the loading jib.

Left: A power shovel used for dinting in soft stone. The forepoling girder in the centre is about to be raised into position under the roof.
Right: An M.R.D.E. Campacker (photographed on the surface) builds packs underground by the thrust of its cam-shaped pushers.

Below: Picks being changed on a Dosco dint-header; it drives through coal at an average of 10 metres a shift or 100 metres a week.

Dinting Machines and Packing Machines

Dinting machines break up the floor where for any reason it is better to do so than to rip the roof. The floor of a coal seam is often soft and there may be occasions when a good roof deteriorates seriously after it is ripped.

Packing machines of various sorts usually include a conveyor that removes the broken rock from the ripping lip and transports it often as much as ten metres into the pack. Sometimes the conveyor also pushes the material tight against the roof to complete the pack.

Full-face Tunnelling Machines

Tunnelling machines (or full-face tunnelling machines) grind away the whole rock face in a similar way to a large diameter oilwell rotary drilling bit boring a hole. The rock is removed from the tunnel face by conveyors built into the machine. There is no shotfiring or shovelling.

In 1882, an early compressed air powered machine, designed by Col. Beaumont of the Royal Engineers, drove a mile or so of tunnel 2.14m (7ft) diameter near Dover, for the proposed Channel Tunnel, and another mile on the French coast. With these machines there was much shovelling even though the excavation was entirely mechanical. But in 1883 further progress was denied by Parliament on the grounds that it represented a threat to the security of the British Isles.

Full-face tunnellers have been proved in many civil engineering projects. They will probably be used for the underground roadways that will have to be driven in the late 1970s and 1980s to create essential

mining capacity. The difficulty in applying civil engineering machines to coalmining is that all electric motors underground have to be flameproof. Flameproof motors are safe in air containing methane but they are heavier and bulkier than other types.

One full-face tunneller used in 1975–76 in Dawdon Colliery, County Durham, was developed by the National Coal Board and Thyssen (Great Britain) Ltd. In normal ground it travelled as fast as had been hoped but was withdrawn in weak ground disturbed by faults. This 680 horsepower Thyssen FLP 35 machine bores a tunnel 3.65m (12ft) diameter. It weighs 98 tonnes and drags many more tons of equipment behind it, including that for erecting the permanent supports, in a train nearly 100 metres long. Mining engineers are, however, worried about the noise from the fans needed to ventilate the cutting tools at the face.

Flameproofness, intrinsic safety, gas, dust and ventilating air
Gas and dust are combated at the face by modern equipment that is of a high standard, often enforced by law, though much N.C.B. equipment is better than the law demands. The Inspectorate lays down standards and if these are not complied with the coalface has to close down or produce less coal, involving higher expenses per ton. Methane is emitted from any known seam in a colliery at a known rate, up to several hundred cubic metres per ton of coal mined. The bulk of this comes out during the coal-getting shifts. Often some of the methane is removed earlier by 'methane drainage holes' drilled ahead

from the two ends of the face. Pipes are sealed into the holes and coupled to exhauster pumps that deliver the methane safely either to the surface or to the upcast air shaft.

Only two types of electrical equipment are allowed underground under British law, that which is flameproof and that which is intrinsically safe. The latter includes low-power equipment such as self-powered telephones or signalling equipment. These must be incapable of generating an 'incendive spark', in other words they must not be able to ignite methane. This is achieved by ensuring that any electrical current passing is too small to make a hot and therefore dangerous spark.

All electrically powered equipment that is not intrinsically safe must be flameproof. This applies to motors, switchgear and any other sources of hot sparks, and all these must be completely surrounded by a strong metal housing. Any joints in the housing for the cover, for example, must be made along 25mm (1in) wide flanges that are machined smooth and bolted in a specific way. The bolts and nuts outside the flanges are shrouded with short stubs of tube welded to the flanges to protect them from falling stones or other impact. The nuts cannot be loosened except by a specially shaped box spanner to guard against attempts to open the unit by an unauthorised person. Only an authorised electrician is allowed to open a flameproof unit, working in specified conditions. For instance the power must be switched off in a way that is known to all concerned and so is safe, since the isolating switch for the equipment to be opened may be 500 metres away from it.

Flameproof equipment is necessarily heavier and bulkier than it would be if it were encased less safely but it does ensure that any

Above: An M.R.D.E. irrigated dust filter extracts dust by dropping water down the filtering screen.

Left: A telephone underground near the coal face; on the right are electric power cables and other equipment in flameproof enclosures.

explosion of gas and air inside the casing does not pass outside it or damage it. Such explosions do occur inside large motors or switchgear but the gases, as they pass out of the enclosure, are cooled by the 25mm wide flanges and the explosion is extinguished.

Motors for modern coalface machines often exceed 100 horsepower. All this power is converted into heat that has to be removed by the ventilating air current. Thus if 200 horsepower of motors is working at full load, this amounts to 150 kilowatts of heat being injected into the air and removed by it.

Enough has been said about gas and heat to show that the ventilating air has to accomplish several tasks, without which underground conditions would be intolerable. But one not yet mentioned, dust removal, is perhaps the most difficult since even the slowest air current lifts dust from the mine floor and suspends it in the air. In practice, the upper limit allowed in the face by law since 30 September 1975 is ordinarily 8 milligrammes of coal dust below 5 microns in size per cubic metre of air. The Inspectorate can close down a face which regularly shows more than this amount of dust.

The dust in the air is limited by law to reduce the hazard to men's health. The N.C.B. have been aware of the danger of pneumoconiosis for many years, have initiated research into it and have produced an instrument – the gravimetric dust sampler – for measuring the dust in the air. It is so advanced that it has even been accepted by the U.S. Bureau of Mines.

The more powerful and effective a coalface machine is, the more gas it releases and the more dust it produces, unless steps have been taken to reduce these dangers to health and safe working. The shearer is a

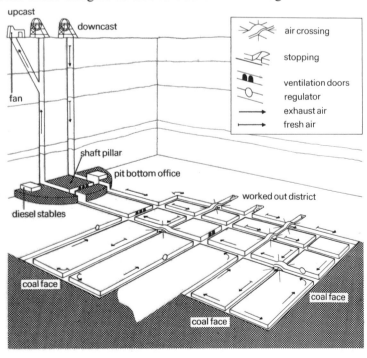

Right: The ventilation and airflow system of a mine.

most efficient machine but it can produce dust. The N.C.B.'s Mining Research and Development Establishment have closely investigated shearer operation for many years and have found that fewer, larger, and sharper picks can reduce the amount of dust formed and increase the average size of the coal; and they use less power. In order to reduce the dust entering the air still further, pick-face flushing was introduced, in which water sprays are ingeniously and precisely directed at the pick points – and at the moment when they are cutting. In this way water is not wasted, nor is excess water spilt on the floor, making conditions uncomfortable for the face workers. The dust is laid before it enters the air stream.

So far as gas is concerned, the pick points are in contact with the coal. It is from the coal that the gas is released, so that the largest concentration of gas is near the pick points. This may lead to a hazardous situation, and ventilation of the picks has been achieved through hollow-shafted shearer drums. A water spray passing through the shaft pulls air with it by a venturi effect and ventilates the picks.

Methane drainage, already described, is a practice used increasingly and is installed solely to promote the safety of underground working. But it now has been also put to commercial use and the methane led out in pipes can be successfully burnt under colliery boilers or profitably sold to large commercial users.

A specialist mining engineer is appointed in every area of the N.C.B., whose sole concern is the underground ventilation, and many individual mines also have their own ventilation officer. He ensures the correct operation of the air circuits, of the methane drainage and so on. He keeps a close watch on the methane contents of the intake and return air flows. If these rise too high or if the air gets too hot, he can advise the colliery manager on how to get more air to the faces concerned without starving other faces of air.

Underground transport
The transport of coal and rock is three-quarters of the work of coal mining, and it is now clear that conveyor belts, where they can be used, are the mining engineer's preference, except in the face. There the all-steel A.F.C. is best. Belts can only be used in straight roads, not too steep (less than about 15° to the horizontal) and not changing from a downward to an upward slope too rapidly. When the slope changes in this way the belt forms a 'valley' shape. If there is no load on it, the belt may then leave its 'idlers', or rollers, and rub against the roof unless the slope changes are kept within limits. The other change in slope, when the belt 'goes over a hill' does not result in the belt leaving the rollers and causes no trouble. These limitations on belt working are not very severe and belts can be used in most conditions that have been planned for beforehand by driving straight roads. Modern belts are even built with steel strands woven into them to transmit the heavy tensions of long uphill pulls.

Because of their reliability and high outputs, belts are being used for hauling the main output of a mine up a drift and can move more coal

Longannet mine in Kincardine has one of the longest conveyor belts in Britain. Here it is seen underground (top) five miles before it emerges at Longannet power station (below).

per hour than any other method of transport now in use, including skip hoisting. A 1.07m (42in) wide rubber belt can move 1,000 tonnes an hour or 20,000 tonnes a day or 5 million tonnes a year, in 250 working days. But even larger belts exist – a 1.37m (54in) wide trunk belt, installed underground at Kellingley Colliery, can move 1,600 tonnes an hour. Such installations may be used at Selby.

At Longannet on the north bank of the Forth, an underground belt 9 kilometres long (5.5 miles) is in regular use carrying more than 2 million tonnes a year of coal out of the mine, mainly uphill. This is a cable belt, in which the pull and weight are carried by steel wire ropes each side of the belt which has two grooves shaped to rest on the ropes. The climb to the surface is about 200 metres (600ft).

Compared with belts, locomotives are much more sensitive to slopes. They should not have to work against a gradient of more than about 1 in 20 and even then the tonnage they can move is severely reduced. Ideally a locomotive track should be sloped at about 1 in 200 in favour of the load, i.e. pulling the full mine cars out downhill and pulling the empties into the face uphill. With such a gradient the locomotive needs to exert about the same pull in either direction. British mines have rarely, if ever, been planned throughout for these conditions but a number of them have good level track near the pit bottom and that is where most locomotives work. Locomotives have the advantage over belts in that they can pass round corners. Most of those used in Britain are diesels but some electric battery locomotives are also used.

Rope haulages are still used in many British mines and for steep drifts (tunnels), where belts cannot be used, are the only possible haulage method. They can negotiate not only almost any curve but also any gradient up to 1 in 1 or even steeper with 'trapped wheel' vehicles, which cannot be derailed. One disadvantage of rope haulages is that the whole length of the road is unsafe for men to walk along when the rope is moving in it. This does not apply either to locomotives or belts.

Ropes are most convenient for awkward roads but their coal throughput is much smaller than that of a belt.

Hydraulic transportation of coal in pipes has been under investigation for many years at the N.C.B.'s Mining Research and Development Establishment but it has the disadvantage of wetting the coal and thus making it difficult to screen and to dry. Nevertheless the work is continuing. Pneumatic transport does not have this disadvantage. It has been found practicable and its present scope is that it could be feasible for improving the throughput of a shaft where the winding

Above: A diesel-driven manriding train.
Below left: A rope haulage system uses endless ropes to pull full mine cars away from the face and empty ones back.

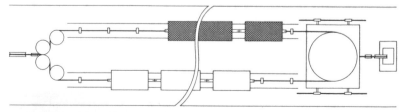

capacity is too small. Alternatively, it might be possible to hoist the output of a remote district through a borehole since an 8in. pipe has been found capable of raising 45 tons per hour up a 1,300ft gradient, using an air pressure of 25lb per square inch. In proportion, a 20in. pipe, with six times the area, might hoist six times as much.

Manriding

The legal length of the underground shift limits the amount of time that men are actually working at the face, and in practice it is often only about five hours because of the time taken to travel between the shaft bottom and the face. Underground, walking is slower and more tiring than on the surface. For these reasons it is, where possible, better for men to be transported to work underground. Belts, rope haulages and locomotives are all used but with strict safety arrangements and at speeds approved by the Inspector of Mines.

Automation and remote control

Automation is not the same as remote control, but in the case of mines remote control from the surface is almost as good as automation. It is now being used for underground and surface conveyors, bunkers, pumps and fans. Mineral winding installations have been automatic since about 1955 in Sweden and there are a number of them in Britain, especially skip hoists.

Below: The nucleonic head (right) attached to the tip of this nucleonic shearer contains a cobalt 60 source to check whether the coal roof thickness is correct, and tilts the shearer up or down as required.

The current tendency to automation and remote control began some years ago and is aimed at eventually fulfilling the mining engineer's dream of manless mining. If the numbers of men in the face in 1947 and 1976 are compared, it will be seen that there are many fewer in the face now although the face outputs are bigger. A face that in 1947 had fifty men getting coal now has only five, but on other shifts in 1947 there were another ten or twenty men cutting the coal, moving the conveyors forward and packing the waste. While manless mining may never be achieved, each little step towards it makes the ton of coal less expensive than it would otherwise be. The full realisation of automation and remote control will not be quick but a real start has been made with certain units and at certain collieries. Most of the work has been done at or inspired by the N.C.B.'s Mining Research and Development Establishment (M.R.D.E.) at Bretby, near Burton-on-Trent. The minute integrated circuits known as 'chips', for example, are so small that they demand tiny quantities of electrical power and so can easily be made intrinsically safe. Such circuits can therefore be used for weekend monitoring underground even with the mains power switched off, by using battery power.

Another example is the nucleonic shearer, of which about ten installations are usually working underground at any time. It is able accurately to control the thickness of the roof coal above it. A cobalt 60 source emits particles towards the roof, and a device in the control head measures the back scatter from the roof. The amounts reflected by coal and stone are different. A small computer in the control head converts an indication of too little or too much back scatter into an appropriate movement of hydraulic rams that tilt the shearer up or down as required. This so-called 'horizon control' or 'vertical steering' has also been achieved by other means.

Powered supports have been remotely operated from the main gate but this is not yet usual. Operation of grouped supports, however, is now being applied.

Cutter-loaders drag a heavy power cable and a water hose behind them. Looking after these while the machine is cutting can occupy all the time of one man but automatic systems have been devised to handle them and are working well.

With tube-bundle techniques, up to nineteen small polythene tubes of 4.3mm bore can be contained in an armoured protective sheath. These tubes can be connected to any point from which gas samples are needed and can bring them up continuously to the control room on the surface. An instrument there makes automatic gas determinations every two minutes or so from the samples and records them. The carbon monoxide content of mine air has been monitored by this method to an accuracy of one part in ten million. This accuracy is advantageous because it can give an early indication of an underground heating before it reaches the dangerous stage of becoming a fire. So far the tubes have been used effectively for distances up to 8km (5 miles).

As in every other large concern, computers are used for storing and displaying information, and the N.C.B. use them for many purposes. They have in fact their own computer company, Computer Power Limited. Stores of mining equipment are monitored by real-time computers, able to indicate the stock of any given machine or component at any moment. Before this could be done, many years had to be spent on the tedious labour of giving identification numbers ('vocabulary numbers') to each of the many thousands of items of equipment.

Left: Tube bundles are used for continuous mine air sampling. Here the tubes are being prepared for leak testing.
Below: Monitoring of the environment: measuring methane in the ventilation.

Above: These boilers, whose fuel is blown in through pipes, are typical of modern coal fuelled plants.

The future

In spite of the greatly increased productivity of the modern miner, labour costs still amount to more than half the cost of production of coal. With an annual turnover of over £2,000 million, this means that the wages bill for British coal exceeds £1,000 million a year. Consequently investigations are continuing, and will continue for the indefinite future, aimed at making this expenditure more productive. Other, parallel investigations will improve the miner's underground environment. Nearly all these investigations are centred at the M.R.D.E. It is interesting that M.R.D.E.'s 'Environment' branch is mirrored in the organisation of the N.C.B. Mining Department.

M.R.D.E. has about a hundred active projects going at any one moment and the list is too long to produce here; but at least their seven headings can be described, with a few examples. Five of these headings are closely related to each other through their dependence on the first one, the coalface. The other two groups are 'basic mining' including metallurgical studies, and a very large group for the testing of equipment, which includes non-destructive testing.

'Coalface' investigations include stable elimination, coalcutting, horizon sensing and the automatic steering that it is intended to provide; as well as the chainless propulsion of face machines, the automatic handling of electric power cable and water hose in the face, and so on.

There is a great variety in roadway drivage machines, that corresponds to the variety that exists in underground conditions. Many machines are being developed or investigated at M.R.D.E. Most of them are used for driving the access roads to coalfaces, where the coal is removed first, and the rock above it later if need be. If the seam is thick enough, e.g. more than 2m (6ft 6in) there may be no need to rip but in Britain such good conditions are now unusual. The heading machine may break out the coal as well as the stone or only the stone, but usually in any case loads out the broken material. Ripping machines do not remove coal but only stone.

So far as full-face tunnelling machines are concerned, the Thyssen FLP 35 drives fast enough in normal rock and produces an excellent, smooth, cylindrical rock surface which is more stable than any surface produced by shotfiring. Other lines of investigation for tunnelling include the conventional metal-mining cycle of drilling, blasting and loading. M.R.D.E. hopes also to develop a more generally adaptable tunnelling machine that will be economic for shorter tunnels than the Thyssen and will cost less.

For underground transport, minicomputers on the surface are being used to control conveyors and to display data about their operation. They have reduced the number of blockages in chutes that are caused by conveyors over-running and have so allowed them to start more quickly, saving money by improving the conveyor running time per shift.

'Coal preparation' improves the marketability of the coal. It produces coals with less ash and less sulphur, but also increases both the

proportion of the tonnage hoisted that is saleable and the financial return to the mine. One indication of its success is that problems are investigated and solutions proposed for important customers including British Steel, the Central Electricity Generating Board, and Associated Portland Cement Manufacturers. A colliery in northern Spain is using three M.R.D.E. Vorsyl separators and M.R.D.E. helped during the commissioning of these excellent, simple coal washers which are also earning foreign currency elsewhere for Britain.

Under 'comprehensive monitoring', a new automatic analysis by minicomputer of the underground environment will monitor data on ventilation, methane emission and methane drainage. Very low level electrical power supplies also have been investigated, and modified, and have received Government certificates of intrinsic safety. Tape recorders for use underground have been accepted as intrinsically safe in this way. Work is also done on sound-level meters, on instruments recording temperature and humidity, on automatic firedamp detectors, oxygen detectors and fire detectors.

'Basic studies' include problems of rock mechanics as well as geophysical investigations for the early detection of seam discontinuities ahead of the face, such as geological faults or old workings. Predictions of conditions in future workings, made by M.R.D.E., include forecasts of the air temperature and of the firedamp emission. Dust control, sampling and noise control are also investigated, as well as metallurgical problems concerning lubrication, wear, cutting tools, mining steels, and corrosion. An engine powered by liquid nitrogen is now being developed at M.R.D.E., that is not only safe and needs no flameproofing, but has the further advantage that it cools the air around it during operation.

Above left: Vorsyl separators (developed at M.R.D.E.) work on the cyclone principle to wash coal.
Above: The M.R.D.E. automatic firedamp detector. It is battery-operated and intrinsically safe, and will eventually supersede flame lamps as detectors.

4. Selby-The Mine
Michael Pollard

The story of the North Yorkshire coalfield is of a steady march eastwards. It began with the people of Roman Britain, who were not enthusiastic coalminers but used coal locally as a fuel where they could extract it without too much effort; accordingly, they dug drifts – sloping tunnels – into the east-facing slopes of the Pennines where there were outcrops of coal near the surface. Sporadic mining continued over the succeeding centuries in the Wakefield area, and even as far east as Pontefract, wherever the coal was fairly easy to get. When, under the patronage of Henry VIII, John Leland made his massive survey of England in the middle of the sixteenth century, he passed through the North Yorkshire coalfield and noted that mining still depended on ease of access. 'The craft', he wrote, 'is to come to it with least pain in the digging.' Further north, incidentally, Leland heard intimations of a problem which was to come to the fore five centuries later at Selby: 'Some suppose that coals lie under the very rocks that the minster close to Durham standeth upon.'

With the coming of the Industrial Revolution, mining in the area between Wakefield, Bradford and Leeds quickly became established. Here, the coalfield was exposed and the seams – including the Barnsley Seam, later to prove of such significance in the Selby project – were thick. Under these conditions, mining brought quick rewards: you exploited the seams near the surface, and the revenue from these enabled you to dig deeper. Improved geological knowledge, drilling techniques and methods of shaft-sinking enabled the exposed field to be exploited to the full, ending, in the east, with a rash of new pits sunk in the Castleford and Pontefract areas in the late nineteenth century.

But coal is a finite resource; the mining engineer must always be moving on. As the coalfield stretched eastwards, the going became harder. The depths of the collieries on the eastern boundary tell the story: at Glasshoughton, near Castleford, sunk in 1870, 440 yards; at Prince of Wales, sunk in 1872 two miles away near Pontefract, 716 yards; at Fryston, sunk in 1875 and later deepened, 546 yards; at Ackton Hall, near Featherstone, sunk in 1888, 576 yards. Wonderful though it was, Victorian technology had its limits, especially when constrained by the limited investment that private coal-owners were able, or prepared, to make. The sinking of new shafts to the east was becoming increasingly costly and increasingly speculative.

The attention of mining engineers therefore turned south-eastwards towards the Doncaster area. As for the area round Selby, stretching northwards to York, logic suggested that the concealed coalfield continued, perhaps eventually joining up with the Durham coalfield; but the problems were known to be immense. The geological structure so clear to the west broke up, and a great cap of Permian limestone, covered with drift deposits, seemed to descend on the coal measures. Even so, in the early years of this century the Earl of Londesborough, the major landowner in the Selby area, commissioned the drilling of a number of deep boreholes in an attempt to find accessible coal. His surveyors drilled at Barlow, south-east of Selby, in 1904. Five years later they tried again closer to the town. In 1913 a third borehole was drilled at Wressle, to the east.

The results were disappointing. The rich seams which had been expected failed to emerge; what little coal was found was in thin seams of poor quality; the geological picture that emerged was even more confused than before. The Earl gave up the quest, and the Selby coal measures remained undisturbed for another fifty years.

By the 1960s, however, a number of things had happened. The coal industry had been nationalised and had begun to look at the hiding-to-nothing economics of working the old pits to exhaustion, winning poorer coal at greatly increasing cost. Changing industrial and domestic lifestyles, placing increasing reliance on electricity rather than primary fuels, had altered the industry's marketing strategy. Drilling technology had developed. Finally, and most convincingly, a borehole at Kellington in the 1950s had revealed, for the first time east of the Great North Road, a number of workable seams of which the Beeston Seam was the most important, but which also included the 4ft thick Warren House Seam, known to be an extension of the Barnsley Seam.

Drilling for coal near Selby: after the drilling rigs (above left and right) come the surface tests (above centre and below left). Their cores (below right) eventually proved that the Barnsley seam was 10ft thick.

Despite the discouraging history of drilling round Selby, the drillers looked eastwards again, and in 1964 began a series of five boreholes: north of Selby, at Kelfield Ridge and Whitemoor; south at Barlow No. 2 and Camblesforth; east at Hemingbrough; all within three to four miles of Selby itself. When the drill-cores were analysed, they showed the Barnsley Seam running towards the east, good and thick: eight feet thick at Kelfield Ridge, nearly 2,400 feet down; four hundred feet deeper at Whitemoor, nearly seven feet thick; and so on. The National Coal Board's regional geologist, R. F. Goossens, put it like this:

'The most important find of this new Selby drilling is the Barnsley/Warren House development. When the Barnsley Seam is traced northwards from Doncaster it splits into the Warren House and Low Barnsley seams, of which the former is the better of the two and the Low Barnsley is sometimes washed out by the sandstone that develops between the two seams . . . To the north of Selby the Warren House and Low Barnsley seams combine to give a Barnsley Seam not unlike that found in the Barnsley district . . . To the east the seam could well continue for some miles east of Whitemoor, but of course will increase in depth to the point where working becomes difficult. To the north

the speculation still exists; a series of large faults could throw the seam out below the Permian, but it seems likely that it stretches for several miles, at least as far as the southern outskirts of the city of York.'

A new coalfield had been discovered; the next step was to 'prove' it. In 1972 a major boring programme was started. On 22 December of that year, the core of a borehole at Cawood, some five miles north-west of Selby, was recovered. It revealed that the Barnsley Seam was, at that point, ten feet thick. Even better, it was 'clean' coal, good enough to send straight to the power stations without washing. 'This', said Bill Forrest, deputy director (mining) of the North Yorkshire Area of the N.C.B. who is in charge of the Selby project, 'was indeed a worthwhile Christmas present.'

Further boreholes rapidly filled in the picture of the new coalfield. The Barnsley Seam was up to eleven feet thick in places. Northwards and eastwards of the area explored, there was likely to be more workable coal; but for the time being it was enough to know that below the Selby area lay about 600 million tons of relatively dirt-free reserves. It was, Mr Forrest said afterwards, 'a milestone in the history of the Yorkshire coalfield'. In fact, it was a milestone for the British energy industry, because, as it turned out, the era of cheap oil was about to end with the Yom Kippur War of Autumn 1973. Of the forty-plus million extra tons of coal a year planned by the National Coal Board, nearly a quarter was there for the digging in the 150 square miles of the Selby coalfield.

Attention now turned to the question of how the coal was to be brought out. As it happened, the necessary new technology – deep-shaft working tactics, highly-mechanised work at the coalface, conveyor systems, improved ventilation, the concept of the miner as a technician – had already been developed at the National Coal Board's newer pits, and especially at Kellingley Colliery, only a few miles away, in the 1950s. Kellingley, opened in 1958, is a high-productivity pit which, thanks to mechanisation, has an output, man for man, about twice that of the older collieries to the west. 'From the engineer's point of view', says Bill Forrest, 'five Kellingleys was the simplest way to tackle the problem.' (Individual deep shaft mines, driven down to some 2,500 feet and taking advantage of what had been learnt and put into practice at Kellingley, would soon have sped the riches of the Barnsley Seam on their way to the markets.) But there were other considerations.

Selby is the centre of an attractive, relatively unspoiled rural area. It lies in flat, open country, claimed to be some of the most productive farmland in the country. The area is not especially appealing in the tourist sense, but it has its own way of life virtually uncorrupted by industrial development. Furthermore, the Ouse and Wharf flow along the eastern edge of the area, much of which is low-lying, and there were likely to be subsidence problems. Selby Abbey itself is an historical monument of some national significance and of even more local importance. And then, above all, there was the example of what the coal industry had done to the country to the south and west, one of

The downcast shaft at Kellingley Colliery, Yorkshire.

the classic areas of irrecoverable industrial spoliation. There, once coal had been found, the grass had turned grey, the landscape had been tortured into grotesque shapes, the base of the local economy had changed virtually overnight, and the quality of life had taken a nose-dive. The National Coal Board was anxious to avoid the errors of the past; but even without the Coal Board's conscience public opinion, heartened by growing environmental awareness and the victories of local community pressure over bureaucracy at Cublington, Stansted and Maplin Sands, could hardly have ignored the eruption of five large collieries in an area which was predominantly rural, and whose inhabitants counted themselves lucky to have escaped the ravages all too evident to the west.

Further exploration in the field provided the solution. On the western edge of the proposed mining area, good-class coal was found at a shallower level. This would enable access to be gained to the coal by a sloping drift connected to an underground road system which would itself lead to the five prime mining areas. Shafts would still be necessary for ventilation and carrying men and materials, but the handling facilities for the coal, traditionally associated with spoil, huge acreages of unsightly buildings and industrial mess, would be concentrated in one place. As a bonus, since the proven coal was so clean – its average ash content was comfortably below the 18–20 per cent acceptable to modern power stations – it could be carried direct from the drift outlets to the power stations without the need for washing facilities, which take up so much room and are the source of so much nuisance at traditional pitheads.

'Literally, the drift site selected itself,' says Bill Forrest. 'Fortunately, at Gascoigne Wood a fifty-acre set of disused sidings exists, and this site coincides with our geological requirements.' Gascoigne Wood, not so much a wood as a patch of scrub, was at the end of an unmarked lane, its grossly under-used siding a relic of the days when it was a marshalling centre for the trains from the coalfields. Dominating the southern horizon was Eggborough power station, with its fellow, Drax, to the south-east. A third power station, Ferrybridge, was a few miles away to the west.

The plan for exploiting the Selby discoveries was beginning to take shape. Mining engineers are not given to hyperbole, but when Bill Forrest talked about Selby his enthusiasm shone through his natural professional reserve.

'In Britain,' he said, 'the opportunity to undertake the design of a completely new mine in virgin territory is very rare. The prospect of such a mine is tremendously exciting to the mining engineer.'

There was considerable excitement, too, throughout the British coal industry. The 1960s had been a decade of debilitation and demoralisation. Large-scale pit closures had weeded out uneconomic and worked-out pits. The average life of a British pit is eighty years, and even after the closures the majority were approaching that age, if not past it. During the 1960s oil had a price advantage over coal as an energy source, and at the same time massive investment in the natural

101

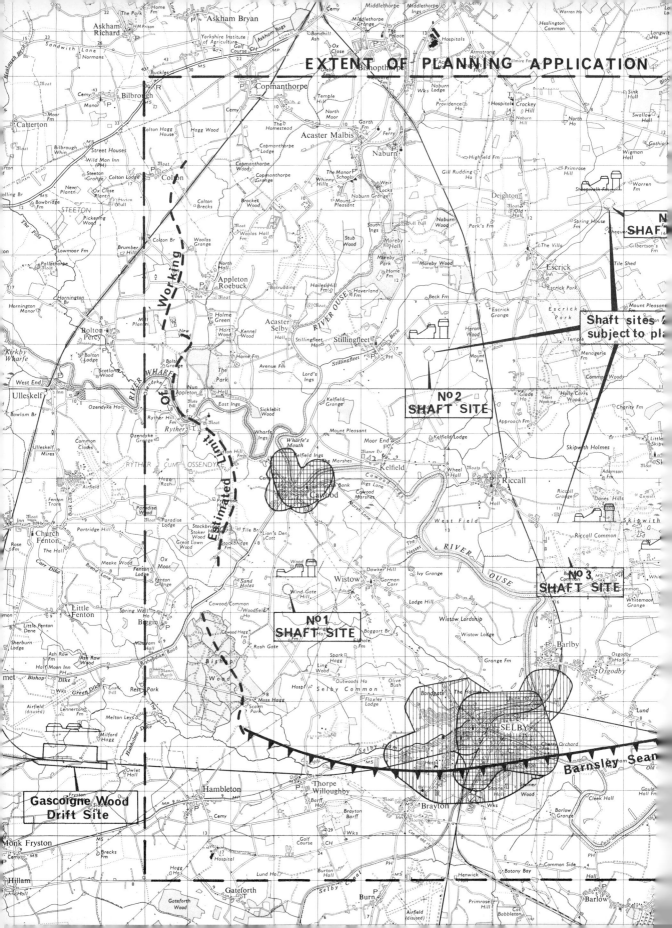

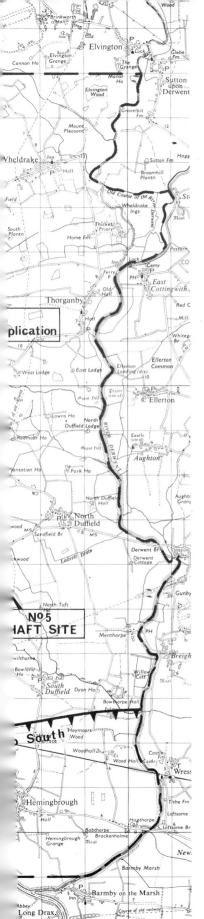

Above: The Gascoigne Wood marshalling yard, six miles west of Selby, where the drift mouth for the Selby mine complex will be sited.
Left: A map showing the extent of the N.C.B.'s planning application (before this was granted in 1976) and the shaft and drift mouth sites. No mining will take place within the areas cross hatched. The areas hatched vertically will be subject to special subsidence control.
Below: Drax power station.

gas programme and in nuclear power made coal look like the poor relation of the energy family. Expenditure on major capital projects fell by 1972 to little more than one tenth of that of ten years before. In line with this, output dropped from 184 million tons in 1960 to about 120 million tons in 1974. But now, just at the moment when Middle Eastern oil had become a political weapon as well as an energy source, and when fears were being expressed about the rate at which all the world's oil reserves were being used up; when earlier estimates of the speed at which nuclear power would develop had been proved over-optimistic; and when economic conditions had given a new emphasis to home production, came the windfall of Selby – an extra ten million tons of coal a year from proven reserves at least until 2015.

The initial plan for developing what might be called the modern wing of the old Yorkshire coalfield is concerned only with the Selby area itself, from Naburn in the north to Barlow in the south, and from Gascoigne Wood in the west to the River Derwent in the east. Although the coal measures may well extend northwards and eastwards at workable depths, any exploitation of these is beyond the present plans. So too is development below the Barnsley Seam, though in fact there is, fifty feet below it, another good seam, the Dunsil, up to six feet thick and capable of being worked from the Barnsley drivages.

But even within the National Coal Board's present self-imposed limits, the Selby project is, as Bill Forrest says, 'possibly the biggest single coal-mining operation in the world'. When full production is reached (according to present estimates, in 1987–88) its five pits will each send two million tons of coal a year through the Gascoigne Wood loading stations, a total production equalling that of all eighteen pits of the N.C.B.'s North Yorkshire Area in 1973. No wonder that words like 'bonanza' are quite uninhibitedly used in the headquarters offices of the Selby project.

The first Selby coal is expected to come out in 1981–82, and from then on production will steadily build up to three million tons two years later, and onwards to the ultimate target. It is a curiosity of modern capital-intensive mining techniques, however, that the major effort comes in the development rather than the production stage, and this is particularly true of working in virgin ground. Development strategies that have been well-planned have a habit of appearing easy in cold print, and in reading the description that follows of the development and production techniques to be employed in the Selby area it is perhaps useful to compare them with the opening of the Kellingley Colliery, scheduled to produce an eventual two million tons of coal per year but at present operating at half that output. Kellingley was established in 1958 and started production seven years and over £15 million later. What is getting under way round Selby is a project five times the size of Kellingley, with the added complications of the greater depth of the seam, a separate drift outlet for the product, and the unknown qualities of virgin territory.

The mining area will be divided into five blocks, each with its own deep shafts for manriding, materials handling and ventilation. The

A model of the Wistow shaft entrance.

absence of facilities for coal handling will enable the shaft sites to be kept comparatively small, about twenty acres each in area. There will be two shafts at each site, each shaft twenty-four feet in diameter. The primary shaft, fitted with a crane capable of lifting up to 16 tons, will be just under 100 feet high and will cater for all normal winding operations for both men and materials as well as taking fresh air to the underground workings. The second shaft, 70 feet high, will fan return air up to the surface and provide standby winding capacity.

Although the mines will have a common outlet for coal through the drift at Gascoigne Wood, each will be independently managed with its own stores, stockyard compound and labour force. By comparison with traditional mines, the Selby pits will be modest employers of labour – perhaps 200 or so men per shift at each. 'With techniques available now', Bill Forrest says, 'it does not seem unreasonable that one man per shift, assisted by patrol men, could supervise all coal-handling from the face to the surface.' Selby is to be a high technology development whose employees will be technicians, not labourers. Automatic monitoring of safety conditions, electronic alarm systems, closed-circuit television and a network of two-way telephones ('despite more exotic innovations, still the most reliable form of communication,' according to Mr Forrest) will make it possible to control operations from a small number of strategic points.

The aim is to schedule the underground works as a whole, driving in along the Barnsley Seam from Gascoigne Wood at the same time as the first shafts are sunk at the selected shaft sites so that the flow of coal out of the drift, rather than out of the shafts, can begin as soon as possible. Inevitably, however, the sinking of the shafts will produce some coal – possibly about 500 tons per day from each site – and this, until the connection is made between shaft bottom and drift, will have to be taken out via the shafts and sent to other collieries for disposal. The first shaft to be sunk will be at Wistow, about three miles north-west of Selby, the first site to be chosen.

The amount of equipment and material needed underground to achieve a production target of two million tons a year per pit is prodigious. Giving evidence at the public inquiry into the Selby project, Bill Forrest gave this picture of the expected underground activity at just one of the sites:

'There will be four coal faces. The total weight of equipment on any such face is about 1,000 tons. This equipment includes 200 powered supports weighing at least four tons each. In addition to these four working faces there will be probably another three faces in the process of being equipped. Besides this there will be a large number of conveyor drives each weighing up to 10–15 tons. Heading machines weighing up to 25 tons each will be used to drive the tunnels. Each mine will operate four mining type locomotives and standby locomotives will be available. Each locomotive will weigh 21 tons . . . Besides this I would estimate that about 80 tons daily of materials such as girders, bricks, cement, rails, timber etc. will need to be transported underground.' The implication of all this is that although the Selby pitheads are to act more as 'service stations' than as production centres they will nevertheless be hives of considerable activity.

Each shaft site will work an area roughly five miles in diameter, using the 'retreat' method. This involves driving roadways into the coalface, one for ingoing and one for outgoing traffic. The face is then opened up between the far ends of the roadways, and the coal is extracted by retreating towards the starting-point. Powered supports, the modern successors to pit-props, hold the roof above the area being worked; as the face is worked, the supports are removed and the roof collapses. The coal is extracted in panels, pillars of untouched coal being left between each panel. The relative widths of panels and pillars can be calculated, in conjunction with geological data, to provide extremely accurate forecasts of surface subsidence. Compared with the hit-or-miss methods of the old days, therefore, the development of a modern underground coal working is a scientific exercise conducted within known tolerances.

The design aim of the Selby layout is to limit surface subsidence to about 0.9 metre, a figure which sounds alarming to the layman but which becomes acceptable when expressed in practical terms, as it was to the public inquiry by the National Coal Board's chief surveyor and minerals manager, R. J. Orchard (see next chapter).

In addition to the careful control of subsidence over the area as a whole, Selby itself, with its eleventh-century abbey, has been given special attention. Not surprisingly, local people were quick to identify the Abbey as one of the possible victims of mining operations, with some gloomsters prophesying that the building would drop by twenty feet, while others foresaw its complete collapse. The worry about subsidence has been allayed by defining an area beneath the Abbey which will not be touched, the so-called 'Selby Pillar'.

It is only natural to be sceptical about statements made by N.C.B. officials at a public inquiry, on whose findings a substantial slice of Coal Board business depends – especially when we live in an age when it is thought advisable to take every public pronouncement with a pinch of salt. But it must be remembered that the National Coal Board is going to be in the Selby area for a long time; that it may well, in the future, want to extend its operations either outwards or downwards; and that in the meantime it is only too well aware that its behaviour and sense of responsibility in the early stages of the Selby development will be watched closely at the time, and possibly be held to account later.

Similar considerations apply to the social effects on the area of the introduction of a new, alien industry. Although, by Yorkshire mining standards, the total labour force of the Selby project, some 4,000, will be comparatively small, it will represent a considerable change in the occupational pattern and tradition in an area with no history of heavy industry. Again conscious of the examples to the west, the Board is determined to avoid the creation of mining ghettoes and to integrate its labour force – some from the declining pits to the west, some from Selby and the surrounding villages – with the existing population, and

A model of Gascoigne Wood drift mouth.

it will be helped in this by the gradual build-up of workers over the eight years' run-up to full production. At the same time, local people – particularly young people – are already being told about the job opportunities that the Selby project will open up for them.

From the environmental and social standpoints, the planning of the five pits, their underground operations, and the recruitment of their employees are critical. The nerve-centre of the whole Selby scheme, the Gascoigne Wood site, is a comparatively straightforward construction job. It is, indeed, the ease of access to the Barnsley Seam at this point that makes the entire project acceptable. From Gascoigne Wood, two central 'spine' roads will be driven below the Barnsley Seam to meet the pits' own roadway systems. Coal coming off the faces will be concentrated on to a belt system which will feed it to underground bunkers, from which it will be transferred to high-speed conveyors with enough capacity to take the whole of the mine's full production, 2,000 tons per hour. Emerging at Gascoigne Wood, the coal will go to a crusher house before being loaded into 'merry-go-round' liner trains which will operate round the clock, to a 23-minute schedule, between Gascoigne Wood and the power stations. The underground system has been designed with built-in spare capacity, and except in abnormal conditions there will be no storage of coal at Gascoigne Wood, though an area will be set aside for emergency stockpiling. As at the shaft sites, operations will be automated, and it is estimated that a total control room staff of forty, working three shifts, will be able to operate the site from its central control room.

When Bill Forrest gave a paper on the Selby project to the Midland Institute of Mining Engineers in 1973, he gave it the title: 'A New Mine for the Eighties'. It's true enough that, in every respect, Selby is a new concept: new in its attention to social and environmental factors; new in the scale of investment involved; new in its emphasis on high mining technology, from geological and drilling techniques to high-speed heading machines and conveyor systems. As Mr Forrest told his colleagues: 'The Selby reserves are a great windfall to the nation and to the coal mining industry... This opportunity will enable the British mining industry to demonstrate to the world that it is now unsurpassed in engineering and mining engineering skills but that at the same time it is able to work sympathetically and in co-operation with all local interests.'

5. Selby – The Community

Jeremy Bugler

Cawood Common was the place where Ronald Goossens, the National Coal Board's regional geologist had the startling confirmation of the real wealth of the Selby field – the core of first-class clean coal 10 foot 3 inches long taken from the exploratory bore hole. Cawood Common, agricultural landscape that had seen slow changes over the centuries, was set for its greatest change when the long core was placed in its long box, and Ronald Goossens hurried back to Doncaster to tell his colleagues what the bore hole had yielded.

The Common, a few miles from Selby, is in the Vale of York which wears the comfortable clothes of fertile soils and prosperous farms. It's flat land, but saved from being tedious by small changes in topography and a weaving of trees and copses in the landscape. A maze of small roads cross and re-cross the Vale, linking with a scatter of small villages which have a quiet air of good living to them. Many of the villages have their own fine churches, usually built of stone and often with a sensible solid tower that you feel expresses the sensible, solid way of life going on below. No one comes to this part of Britain to catch his breath at the classic beauty of the English landscape, but every week of the year has moments when the light on the fields and trees of the Vale of York would make even a dull man pause and look.

Only in concentrated areas or on the fringes has the predominantly rural air of the proposed new coalfield been marked by industrial development. To the south, the huge chimney stacks and waisted cooling towers of the massive power stations of Drax and Eggborough are visible. In Selby town itself, quite a variety of industrial activity takes place. Men and women work in the huge stark flour mills of Rank Hovis McDougall on the river banks of the Ouse; in the animal feedstuff works of BOCM Silcock; in pickle factories and bacon curers and other works. Selby is a town that already offers a good range of jobs to its people, a town that feels itself of some importance, with the London to Edinburgh trains stopping at its station and the tourists pouring in at summer time to look at the new-scrubbed medieval Abbey. For all that, Selby is fairly compact and you need only take a few turns of the steering wheel to be on the road to Cawood Common and the biggest business of the area: the land.

Very naturally, Ronald Goossen's long core of coal produced a shock-wave in the district, some shocks of pain and some shocks of pleasure, but shocks nonetheless. The realisation that your whole way of life, be it farming or flour-milling, rests on a thick Barnsley coal

Right: The town of Selby with the River Ouse flowing through it and Selby Abbey (centre), looking northwest.

Below: The River Ouse crosses the low-lying flat land of the Vale of York, here by the Ferryboat Inn, Thorganby.

seam which will inevitably become the dominant industrial activity of the area – that realisation stops most people in mid-thought. The news that the local people were sitting on the richest coalfield in Britain, destined to be the most productive coalmine in Europe, very understandably stirred up a host of fears, many of them real enough.

For some, the coming of a mine meant the coming of miners, and that they feared. A Selby councillor who started a fund to fight off the mine said that he had to tell the local people what mining communities were like. He planned to show films and colour slides of other mining communities. 'We aren't passing judgement,' he said. 'We will simply be saying, "Is this for you?" to the local people.' The *Yorkshire Post* reported that some local people were 'xenophobic' about miners, though adding, 'The image of miners as a race apart is propagated chiefly among the "outsiders" who now often outnumber the indigenous farming community.' The influx of 4,000 miners plus their families *is* a large amount of new comers, and anxiety by the local people that this 'invasion' be properly handled was both right and proper. That was one fear that had to be met.

Another fear, again with a real basis, lies in the nature of this flat land. The fields in their distinctive pattern in the Vale of York are crisscrossed by drains and dykes to try to dry the fields early in the Spring. The water table lies very close to the surface in Selby country; in the parlance of one man: 'The land here floods like the clappers.' Indeed, one of the ways in which the land sometimes interrupts its flatness is the raised-up river banks, work that people have done to keep the land from flooding. The River Ouse system, though, still breaks through its banks in extreme conditions. Most of the local people remember the year of 1947, after the hard winter, when the Ouse flowed over its banks, covered the fields and flooded houses up to their bedrooms. People who live in flood-threatened land are nervous about anything that could upset the subtle balance of the way the water is drained away; the more knowledgeable of them also realise that it is over sixty years since the coal mining industry was set the problem of mining under a low-lying flat area of land like Selby. A lot of knowledge has come aboard since then, but the nervousness is understandable.

It centres on subsidence, which was and is perhaps the chief physical fear that any community faced with a coal mining project experiences. The local farmers had all read countless stories in the *Farmers' Weekly* over the years of fine fields that had been turned into unproductive marshes because of mining subsidence – especially dating from the times when the private mine owners were not obliged or moved enough to pay for restoring land that they had ruined. The Selby area farmers wanted to know what was going to happen, how their cropping might be disturbed, and most important, how they would be paid for damage. Anyone living near a river bank which might be subsided was especially worried.

Subsidence was also the epi-centre of the fears of people with buildings, from the owners of the semi-detached houses through to the

farmers with large barns, market gardeners with long glasshouses to owners of industrial premises of a huge square footage. Everyone in Britain has seen pictures of buildings leaning over at a crazy angle because someone's pulled the rug out from under it. Vicars were worried about their churches – especially the Reverend J. A. P. Kent, Vicar of Selby Abbey.

People feared, too, the coming of the traditional impedimenta of the mine: the winding towers, the slag heaps. They feared the intrusion into the rhythm of their life of the mine-constructors: the lorries that would have to be used to take away the spoil and, in the early stages, the coal itself. In the narrow lanes that link up the little villages of the Vale, a car coming the other direction is often enough to force drivers to go up on the verges to get past; what would it be like with huge coal lorries, many to the hour? People, especially the older people, feared the noise and dirt a mine might bring.

No one makes any bones about the change that faces the people in the Selby coalfield area: a huge change to livelihoods and landscapes. As Coal Board men said openly: 'You can't put a ten-million-ton-a-year mine down and nobody know that it's there.' In terms of inevitable disruption, then, the National Coal Board had a task on its hands to win at least the support of some part of the communities and the dampened-down hostility of other parts. The National Coal Board, however, also had factors massively on its side which have made a huge difference in the attitude of the community to the project.

The events that made this difference occurred, or started to, three months before the Cawood Common core came to the surface. Following the outbreak of the Yom Kippur war, in a few weeks the flow of oil to the West was cut back and the prices shoved up, almost double. In January 1974, in defiance of the predictions of most economists, oil prices again more than doubled. Western nations were faced with huge bills for importing oil, and Britain's coal reserves received a flush of new Government interest. The calm assumptions of government after government that our national energy needs would be met reliably and cheaply by imported oil were thrown out neck and crop; even with the North Sea oil reserves proving monthly larger than first predicted, coal was needed like it had not been needed for twenty years. Developing the 110 square mile Selby coalfield in particular would save the nation £250 million a year to the balance of payments.

The nation's need has been a crucial factor in the wide acceptance of the local community that the mine was going to have to come. Resistance to the mine was largely, though not entirely, shifted on to how to make it as good a neighbour as a mine can be. Contrast this acceptance with the opposition that Rio Tinto Zinc faced in the early 1970s when they applied for planning permission to drill for copper in the Snowdonia National Park. By no reasonable stretch of imagination could the local people see that Rio Tinto Zinc's work had a sense of national urgency to it; it was rather a large mining corporation seeking to find out reserves for exploration a long time in the future.

111

They were faced with massive opposition at the public inquiry; they were harried by a storm of critical articles in the newspapers and astringent television programmes. What was good for Rio Tinto Zinc was not thought necessarily good for Britain.

The National Coal Board, in contrast, were being asked by the Government to hasten to the task. The Selby project grew with the national pressure; from a 2 million ton-a-year mine to a 10 million ton-a-year production figure. By 1974, the broad N.C.B. plans were clear in their application for planning permission. The Board wanted permission to mine coal over an area of 71,200 acres (29,000 hectares). They conceived one drift mine and five satellite shafts. The coal, apart from the early stages when roadways were being driven from the five shafts, would be taken out of the drift mine and sent by rail to one of the Central Electricity Generating Board's power stations.

Even before 1974, the N.C.B. had made decisions about their approach to getting the community's acceptance and co-operation. In one scene, it was all summed up when Peter Walker, then Secretary of State for Industry, and Derek Ezra, N.C.B. chairman, stood in the yard of the Kellingley mine in Yorkshire and declared to television cameras that the Selby mine would be 'clean'. It would not be associated with the detritus, dereliction and despoliation of the coalmine of George Orwell's day or even of post-war days. Derek Ezra was interviewed by the *Yorkshire Post* in July 1973, and declared that Selby would be different: it would have no conventional headgear, no atmospheric pollution, no preparation plant, no heavy road traffic, no rail sidings.

As it happened, not all these expectations could be fulfilled, but Derek Ezra's statements do show that in the very early days of the Selby project, the N.C.B. set itself to see if it could build 'an environmental mine', minimising the damage to the surroundings. To this end, it adopted a reasonably open attitude to telling the local communities what they were doing. 'We felt we should inform and co-operate with people as much as possible,' said Bill Forrest, the man with overall responsibility for the project. 'Our attitude didn't come from a specific decision taken round a table; we just wanted to take this approach.'

Accordingly, right at the start, the N.C.B. started a programme of letting people know what they were up to. Bill Forrest and Fred Sanderson, the Yorkshire coalfield's chief public relations officer, began a tour of local communities, speaking to groups and explaining what the N.C.B. was planning. Many people in the Selby area got a copy of *Selby Newsletter* – a broadsheet from the N.C.B. setting out its position: what it had found, what the procedure was, and explaining that the new miners would be 'integrated not segregated' into communities; it set out some facts about mining subsidence. The first issue of the *Selby Newsletter* is dated 1973, which shows that the 'open' approach was N.C.B. policy from the first days.

Such an approach has its pros and cons. It undoubtedly caused the N.C.B. to 'think aloud' about its plans before they were really tested,

The *Selby Newsletter* was started by the N.C.B. in 1973 to help inform the people in the Selby area of the facts and figures of the Selby project as planning progresses.

proved and finalised. This had the effect of raising false hopes and laying the N.C.B. open to accusations that it was 'soft-soaping' the public: getting public approval for an 'environmental' mine, and then – later – announcing the unpleasant details of the mine. The key issue here was the height of the winding towers. Senior Coal Board staff openly said that they hoped the winding towers would be only 40 feet or 12 metres high – little more than the height of a typical house in the district, and easy to screen, factors of great importance in the flat landscape of the Vale of York. The N.C.B. hoped that they could use winch-gear instead of winding wheels for raising and lowering the cages in the shafts, and the first N.C.B. calculations showed that they were right. Later, however, new facts came to light which, the N.C.B. says, forced them to return to the technological orthodoxy of the tall winding towers – 96 feet high. Quite a difference, although it should be said that this height is much less than at existing collieries of comparable size.

Local people, some of them, accuse the N.C.B. of 'manipulation'. The N.C.B.'s mistake, however, may also be seen as part and parcel of trying to tell people what they were planning as the project developed. When Fred Sanderson went to talk to groups of people in village halls, taking with him a film the N.C.B. had made about the project, his behaviour was in some contrast to that of the public relations officers of other large industrial concerns about to start a major project: they release as little information as possible. In retrospect, of course, the N.C.B. could be said to be wrong to raise people's hopes before the Board had really done its sums. You shouldn't make definite declarations unless you're definite. On the other hand, the N.C.B.'s fault can be seen as being born of good intentions.

If the N.C.B. had had to default on some important statements, it has stuck by others that contribute to making the mining more acceptable. Once the N.C.B. were set to produce 10 million tons a year of coal out of the 110 square miles of the Selby coalfield, the easiest engineering solution (as was discussed in the previous chapter) would be to sink five modern but conventional deep shaft mines, each a separate complex. 'Five Kellingleys' was the way the N.C.B. thought in the first stages of the project, but once the realisation of environmental considerations struck home, a rethink started. Five separate mines scattered around the Vale of York would have done real violence to the Vale of York. The Board had anyway proved to itself that 'environmentally clean' mines were not impossible. At Longannet in Scotland, two new drift mines have been successfully created and blended into the local landscape (made easier by the rolling nature of that landscape).

When a drift mine was considered, theoretical environmental advantages were immediately apparent. The need for five Kellingleys, each producing coal at its own pit head, was done away with. Whatever the disappointments in the future over some N.C.B. hopes, the concept of the drift mine, with all the coal taken away at one rail head, is undeniably a step towards minimising the damage to the countryside.

Through 1973 and 1974, people of the Selby area were seldom unaware of the plans that were developing to change their surroundings. Quite early on, clear differences of opinion were discernible among the community. A Selby councillor pledged himself to form a Selby Protection Club, with a membership fee of £1 and a hoped-for membership of 25,000. The councillor pledged that he would 'fight tooth and nail against the rape of our countryside.' Other groups were vocal in favour of the mine. The Selby Trades Council and the Selby Chamber of Trade were among these, with Mr Peter Roff, chairman of the Chamber of Trade, saying, 'A coalmine will be a gold mine.'

All these groups and counter-groups, fors and agins, had their full say when the official public inquiry was held in the Spring of 1975 into the N.C.B.'s plans. There is no better way of understanding the community's reaction to Selby than to follow the hearings in the little Museum Hall of Selby.

The public inquiry in Britain is the 'courtroom' of planning. Any major new development or any controversial scheme is likely to end up in a public inquiry. After the organisation concerned has put in a planning application, it is normal procedure for the Department of the Environment to 'call in' the application and to make it the subject of a public inquiry. The inquiry is presided over, in England and Wales, by an Inspector, usually a civil servant appointed by the Secretary of State for the Environment. He often is assisted by specialist assessors. He listens to both the applicants and to all the objectors. He records also every word put to him by the planning authority and other involved local authorities; by experts of every shade and hue; by organisations from associations of employers to groups of butterfly hunters. He listens, asks pertinent questions, and, when everyone has had their say, retires (usually back to London) to write up his report. In practice, it takes an inspector about one day to write the report for every day of the public inquiry; in time, his report, usually with a recommendation, is sent to the Secretary of State, who can then either accept or reject his recommendation, or put in conditions on his approval of the project, or other terms. Whatever the Secretary of State does, he must publish the report of the Public Inquiry.

Inevitably, the format of the inquiry makes it something of a contest, a struggle between aggressors and aggrieved, between preservations and constructionists. As such an arena it has severe limitations from the planning point of view. Every planning purist's ideal solution is a machinery that has never once been used since it was established under the 1968 Planning Act: a 'Planning Inquiry Commission'. Such a Commission has much wider powers once it has been set up by the Secretary of State. It can call its own witnesses, do its own research and examine all the broader implications of a project while doing without the lawyers' interjections and the tit-for-tat of the public inquiry, which may go on a very different path than the Search for the Truth. However, the Secretary of State did not call for a Planning Inquiry Commission and so, in its stead, Selby got its public inquiry.

Subsidence can be carefully estimated and controlled in modern coal mines by planning the thickness of coal to be extracted and the proportions to be worked. The surface effect can be calculated within millimetres, as in the hypothetical diagram (right).

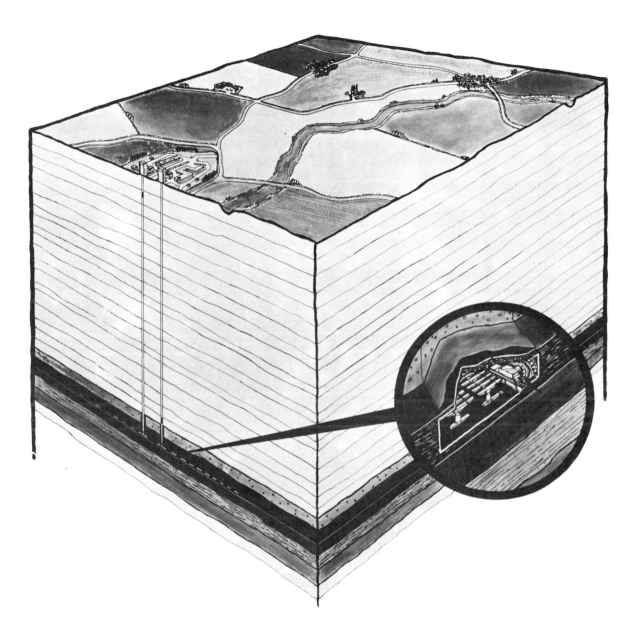

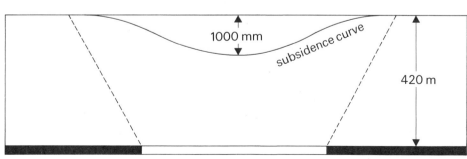

On 2 April the Inquiry opened. Mr Matthew Adamson was the inspector and Professor E. L. J. Potts of Newcastle University was the assessor on the mining evidence.

There were massed ranks of witnesses for both sides, and great public interest. The N.C.B., apart from its legal advisers, had fourteen witnesses; North Yorkshire County Council (the planning authority) expected six witnesses, Selby District Council ten witnesses. There were to be others from the West Yorkshire Metropolitan County Council – worried about the mine's effects on its own area; the Yorkshire Water Authority – worried about drainage and flooding; British Waterways; British Railways; the Yorkshire and Humberside Council for the Environment; the Yorkshire Naturalists Trust; the National Farmers Union and the Nature Conservancy. In addition there were spokesmen for all the Parish Councils of the affected villages and dozens of individuals who wanted to make their views known.

Opening evidence for the N.C.B., their legal counsel, Mr John Griffiths, Q.C. said, 'This is probably one of the most important planning inquiries to be held in this country.' As the inquiry developed and the N.C.B. made its case and other witnesses began to speak, one issue began to underpin a whole host of fears, rather as the coal holds up Selby. And that issue was subsidence.

How much the land would be subsided when the coal was taken out was explained early to the inquiry by a world authority on subsidence, the N.C.B.'s chief surveyor and minerals manager, Mr Eric Orchard. In the bad old days, coal companies took out the coal and the land fell with a thump, collapsing terraces of housing and flooding fields. Since then, something of an art and science of subsidence, not always exact perhaps, has been developed. Orchard explained to the inquiry how the design of mining layouts has been perfected so that a lot of coal can be taken out without complete chaos on the surface, indeed with minimal damage to buildings on the surface. For Selby, especially with its drainage problems in mind, a limit of 0.9 of a metre had been set for subsidence, and the N.C.B. planned this level in view of their proposal to extract the 2.4 metres of coal, which in most part of the coalfield is the maximum that can be extracted.

Orchard set out how this could be done: how the amount of subsidence caused by working a longwall face will be determined by the width of the coal extracted, the length for which it is extracted and the thickness extracted. By leaving pillars of unworked coal at intervals, the subsidence would be strictly controlled. At Selby, the maximum of 0.9 of a metre of subsidence would be produced on the surface by working panels of coal 100 metres wide with pillars 35 metres wide where the coal was shallowest, or panels 320 metres wide and pillars 110 metres wide where the coal was deepest. Subsidence was now a controlled science, not a question of 'extract, grabbit, and run'.

What happens to buildings on the surface when the land is subsided was also explained by Orchard. Buildings got damaged when the curvature in the ground's surface as the coal was removed caused the surface to expand or contract horizontally. Long buildings are put under most strain and semi-detacheds are least at risk. Orchard said there would be very little damage indeed to houses and farm buildings and even long buildings would not be severely damaged. A 200 foot long building, for example, would be subjected to a strain of one millimetre in a metre; a few small fractures or one larger fracture might show on the outside of the building but this, he said, could be repaired without difficulty, and the N.C.B. would be paying for the repairs.

Reassuring as Orchard's evidence is, many of the Selby community naturally are still very worried about subsidence. Take the Vale's medieval churches, for example. Many of the Vale's villages have fine churches, excellent historic buildings that in a country less prodigously provided with old churches would be tourist attractions. The N.C.B. clearly have a duty to protect such buildings; they believe they can do so without leaving a pillar of unmined coal beneath each church, an approach that would make the Selby mining enormously complicated as well as taking coal out of production. The N.C.B. in the past have found that damage to old churches can be minimised by precautions such as taking out trenches around the building. These trenches are either left open or filled with a compressible material so that ground pressures – the push-and-pull – are reduced.

One church in particular, though, in the area could not be treated in such a manner, and this church became both a test of the N.C.B.'s profession that it was concerned about the local environment, and also the key to a whole series of agreements made at the public inquiry between the Coal Board and various local interests.

The church, of course, is Selby's Abbey, a marvellous building with a superb heavy arched Norman nave, a jewel of early English ecclesiastical architecture. It is also a very large building indeed, much too large to be accommodated by a ring of trenches to take ground pressure. Serious structural damage to Selby Abbey was possible if the coal beneath was mined, and in addition, the Abbey itself is surrounded by a conservation area, including many local buildings and streets. Further, the Trustees of Selby Abbey began, very properly, to be sensitive about what would happen to the Abbey with the coming of the coalfield. The Trustees commissioned an architect to investigate the Abbey foundations, and he produced a report which indicated that the water table under the Abbey has remained stable about four feet down ever since the Abbey was built.

This apparently ordinary finding has important implications. The architect showed that no burials had taken place below four feet down, and that the original foundations of the Abbey employed mortar near the surface, but no mortar four feet down and below – the implication being that the ground was too wet there for mortar. The water table, therefore, has been constant below Selby for hundreds of years, and it's quite possible that any change in the water table – possible with mining – would threaten the Abbey in some way. No one knows what happens to foundations when they dry out after years of being submerged; the clay base beneath the Abbey, says the Reverend Kent, could dry out or shrink and there are wood foundations in some places which would crumble on being dried.

In the first days of the coalfield planning, the coal board experts accepted that they would have to leave a pillar of coal beneath the Abbey. The risks otherwise were too great. So it was that early in the days of the public inquiry, the N.C.B. and Selby Trustees reached agreement on how to protect the Abbey. The N.C.B. agreed to leave a supporting pillar of coal, and also undertook to be responsible for any damage attributable to mining. The Reverend Kent sees it thus: 'If the water table should drop, and the drop was clearly due to mining, the N.C.B. will either repair the Abbey or pay for preventive action if the Abbey is damaged by the drop in the water table. In our view, if the water table drops after 2,000 years, it will take a lot of laughing-off by the Coal Board, and we are satisfied at the safeguards in our agreement with them.'

Before the inquiry, the Selby Abbey pillar was enlarged to the Water Authority pillar which would protect the Ouse flowing through

Selby and coincidentally part of the town itself. This in turn was extended to cover several large industrial premises in Selby.

The Selby pillar, however, soon became a matter of great debate. Not unnaturally, most people in Selby wanted as large a pillar beneath them as they could get, for security against subsidence. The Selby District Council, the local authority, were worried about the town's sewers. These were laid first in the latter part of the nineteenth century, with more recent additions radiating out to the newer parts of the town. Wouldn't these old brick pipes crack, leak and back-flow in the event of subsidence? Selby's Engineer Peter Renton feared so, especially many of the trunk sewers of the town, which would be very expensive to repair.

So it was that the Selby council was seeking a much larger pillar of coal beneath the whole town. The Abbey had declared itself satisfied and dropped its objections to the N.C.B.'s planning application. A number of firms in the town followed suit after they got protection agreements with the N.C.B. – firms like BOCM Silcock, Allied Mills, the Danish Bacon Co. Ltd., RHM Flour Mills, Yorkshire Chemicals,

John Rostron & Sons Ltd., John E. Sturge Ltd. and Cochrane & Sons. But both the Selby Council and the North Yorkshire County Council held out at the inquiry for more protection. County Planner R. Cooper Kenyon argued that, with the N.C.B.'s pillar, a number of industrial premises could still be affected by subsidence even though they were not being directly undermined; that Selby has many streets with long terraces of houses that could be damaged, and that there are a clutch of 'important buildings such as schools, hospitals and churches which could also suffer severely from subsidence.' The County Council therefore argued for a much larger pillar, and for the protected area to be enlarged to include the whole part of the built-up town of Selby. At the inquiry and afterwards, the N.C.B. couldn't agree to this larger pillar. By some calculations, it seemed as if it would entail £84 million in lost coal extraction. The issue was left for the Secretary of State to decide, a proper thing since the N.C.B. found it difficult to leave so much coal alone and the local authorities found it difficult to take any risks over subsidence. The Government, then, decides the size of Selby's pillar.

Selby was not the only place where the N.C.B. became committed to leaving coal unmined because of the fear of flooding. Another place was Cawood where a particular bend in the river needs protection – a point that was driven home to the Coal Board when they visited Cawood. The local people, no fools at public relations, took the

The town and fields of Cawood, by the River Ouse, looking north.

N.C.B. men to the pub – and pointed out the line high on the wall where the water had reached in 1947.

Flooding caused by subsidence is one of the chief anxieties of the Selby area people, and no section more than the farmers. As one farmer, Makin Bramley, put it: 'If we drop this area three feet, we will be turning an agricultural area into a duckpond.' The position of the farmer was put with great succinctness at the public inquiry by another individual farmer, Francis Outhwaite of Turnhead Farm, Barlby, near Selby. He and his son John Outhwaite farm 245 acres that at the turn of the century was about the worst land a man could be stuck with – infertile and ill-drained. Today, about half the farm is classified as the very best agricultural land and 90 per cent of it comes within the top three Ministry of Agriculture grades. And the reason for the improvement is largely the drains. The story of the Outhwaite farm is one of constant improvement of the drainage, and any farmer will tell you that without good drainage you cannot have good farming. Fields in the area are drained by dikes and land drains, which obviously could be disrupted by subsidence. Francis Outhwaite's farm actually extends along the River Ouse, so that it is prone to sudden flooding. The Outhwaites and other farmers fear the sight of drowned fields and the loss of farm income not just for one year but for a number of years into the future, for the process of re-draining fields can be slow.

The farmers' case was put again by witnesses for the National Farmers' Union, who pointed out also that almost exactly half the land in the Selby area is in the top two grades – the sort of land that Britain depends on for its food production.

The Coal Board never attempted to pretend that damage to the intricate draining of the fields was not going to happen. They admitted it at the inquiry and promised that the N.C.B. will do what it can to put a field right after damage, to repair or pay compensation to damaged farm buildings, and to pay compensation for the current year's loss of crop. Crucially, the N.C.B. also agreed to make *ex gratia* payments for crop losses in any subsequent year 'until such time as the damage is made good or the Board elects to make a lump-sum capital payment of any permanent depreciation of the land.' And any permanent loss to the value of the land itself (and not just its crops) because of subsidence would meet with compensation.

No one can represent the Vale of York's farmers at being delighted at the coming of the coalmine. Today most of them probably accept it as a necessary evil, interfering with the essential business of life: producing food. However, if the N.F.U. is anything to go by, they are content enough at the undertakings and the guarantees that the Coal Board has given them.

But what of other people – the ordinary householders and shopkeepers, and not just the farmers? Their position was made clear at the inquiry by the evidence of the N.C.B.'s Director of Estates, Alan Dickie. He told them how once a householder suspects that his dwelling is damaged, he notifies the Coal Board, whose officials then meet with the owner and decide whether or not the damage is due to subsidence caused by coal workings. In cases where it is likely but not certain that subsidence is the cause, the onus is on the N.C.B. to show that the damage is not subsidence damage. The N.C.B. have to decide whether to repair the damaged property or to make a payment instead; sometimes, they do a bit of both. It isn't very often that a house is so damaged by subsidence that it's actually uninhabitable, but if this does happen the Board has to temporarily house the dweller while his house is being repaired, and in the very rare cases where it is irrepairable, it must pay him the full value of his house plus a compensation sum for being disturbed. Such drastic action and trouble, said Mr Dickie, was very unlikely to occur at Selby; it is more present in the older coalfields where a number of shallow seams have been worked.

Subsidence can also cause damage to unlikely places. A possibility that the N.C.B. had to consider in the Vale of York is Skipwith Common, for this wet heathland is the largest single tract of such land in the north of England. With botanical abundancies like the Marsh Gentian, it is also particularly important for its bird and insect life. It has about eighty-three species of birds breeding in it, a very high number, and large numbers of wildfowl use it. More than a third of the 900 species of British butterflies and moths have been recorded at Skipwith – some of them very rare indeed in Britain. And its importance has been recognised: it is classified as a Site of Special Scientific Interest under the 1949 National Parks Act.

Naturalists are very reasonably worried that mining could destroy the delicate drainage and thus the ecosystem of the common. Accordingly, the Nature Conservancy sought from the Coal Board a number of conditions and assurances to protect it – particularly surveys before and during mining and a commitment to do remedial works to restore drainage patterns should they be changed.

Not all the Selby problems could be a matter of such agreement as Skipwith Common. Some problems are much more intractable, and this is partly the case with the matter of transport: how the coal is to be carried from the mine. In the long term, there is little problem: the coal will be taken by rail from the one point, at Gascoigne Wood, to the C.E.G.B. power station. It is the best way, environmentally and economically, to do the job. The problems are tied in with the build-up and creation of the mine. The separate shafts have to be sunk, and the waste so created has to be carried away – by lorry. Once the shafts have been sunk, until the drift and shafts are connected, the dirt from the shaft end of the drivage has to come up the shafts and be carried away – by lorry. (As soon as they are connected, all the coal will be taken out at Gascoigne Wood and put on the rail line.) There is no way round this, but it does mean that some small villages will have heavy lorries on their local roads for some months as the mine is constructed. For example, about 500 tons a day of material will have to be taken out of the Wistow shaft for a matter of nineteen months; another shaft will be used in this way for three years. It's a transient

The church and village of Riccall, near which one of the five shafts will be sited.

problem, but given the necessary size of the lorries, one that is bound to cause some discomfort to residents. The N.C.B. has given undertakings to make sure that wheel-washing plants will be installed at all the shafts so that muck and mud is not put all over the road, but, at the end of the day, a lorry is a lorry is a lorry.

As the inquiry went on, transport assumed more and more significance. The people of Selby were shaken when British Rail gave evidence and said they would have to re-route the railway line because of the coalfield. This meant that Selby would no longer be on the London–Edinburgh line. Some Selby people felt the town had lost a major transport link and was being down-scaled in importance. Given the amount of direct Selby to London or Selby to Edinburgh passengers (or vice versa) the damage is probably more psychological than real to the town's economy, but understandable nonetheless.

Real trouble, however, undoubtedly came the way of the N.C.B. in the matter of the winding towers for the satellite shafts. Despite the earlier hopes of men like Derek Ezra and William Forrest, the N.C.B. came to the inquiry and said that technological reasons had forced them to ask for planning permission for winding towers at each shaft of 96 feet 8 inches and of 70 feet. These heights were undoubtedly a shock to the local people, and it mattered little that towers of 140 feet and even 180 feet were typical at other large collieries. They had come to expect an environmental mine and some objectors argued, not unreasonably, that the most unpleasant part of mines, bar the slag heaps of old, is the winding tower. Surely modern technology could solve the problem.

The Coal Board official who got up to explain the height of the towers was James Blelloch, the N.C.B.'s Director of Engineering. He said that the original heights indicated by the N.C.B. – 40 feet and 70 feet – were made when it was not appreciated by the Board that heavy equipment and materials would have to be taken down the shafts, and not just down the drift. Lower towers would have been possible if the shafts were to be used for manriding and light materials only. After the careful calculations were done, though, winding gear capable of handling 16 tons was necessary at each shaft, and this gear made 96 feet 8 inch towers necessary. Mr Blelloch had a detailed breakdown of each foot of each tower, to show why the height was, in his view, necessary.

Residents may have been impressed but their opposition to the height of the towers remained, and it cannot be denied that in the flat landscape of the Vale, the towers will be 'intrusive', as Selby District Council's planner termed them. Residents staged some demonstrations – the people of the little villages of Escrick and Deighton flew balloons to show the height of the towers. This local residents' association argued that the N.C.B. was getting 'a mine on the cheap'. Its chairman, Maurice Vassie, asked 'When tens of millions of pounds can be spent re-routing motorways on environmental grounds, why cannot a fraction of this sum be spent to prevent the industrialisation of the rural landscape above the Selby mine?'

Maurice Vassie argued that the towers should be 'counter-sunk': that a large hole, crudely speaking, should be excavated and the towers built in the holes. The N.C.B. opposed this idea strongly, arguing that it would cause problems in several areas: notably efficient operation and safety. Neither side gave ground; and again perhaps the Secretary of State is in the most neutral position to judge.

On the question of where the shafts were to be sunk, the N.C.B. found itself able to be more obliging. It was able to negotiate with local parish councils, and the Q.C. for the parishes and Selby District Council was able to tell the inquiry that agreement had been reached on some of the sites. In some cases, he said, the new siting involved some loss of convenience and coal, and he was obliged to the Board. Thus the Stillingfleet shafts were shifted to a point below a ridge to the east of the village and not on the ridge itself as first proposed; the Riccall shafts were shifted to a site south of the original point; the Escrick shafts moved to the east of the original site; and the North Duffield shafts were put as close as possible to the junction of the A163 with the Skipwith–Cliffe road. Parishioners here, while still unhappy about the arrival of the shafts, perhaps tended to feel that they were being positioned in places where they would not be the greatest eyesores.

One parish, though, was not at all happy. The No. 1 shaft at Wistow was positioned at a site that thoroughly displeased the local people, and they maintained their objection to the site consistently, and reached no agreement with the N.C.B. Cawood Parish Council backed up Wistow, too, in its objections, and its desire to move the shaft nearer to Bishop's Wood, where it felt that good agricultural land would be safe and roadworks reduced. The Long Lane site at Wistow was objectionable. And here there was no agreement.

As the public inquiry ran, more and more agreements were made between the N.C.B. and affected parties. In this light, the Selby public inquiry was interesting: it assumed a creative character, securing agreements and patching differences. It also brought up some ideas, such as one proposal from an Escrick resident. His name is Dr Michael Chadwick, a senior lecturer in biology at the University of York. Like a good article in a newspaper, his evidence concentrated on making one point well. He proposed that a watchdog body should be set up to monitor the progress of the mine. He termed this body 'an Environmental Consultative and Liaison Committee' and suggested it would have an independent chairman and representatives from the local councils, the N.C.B., the Yorkshire Water Authority, the Nature Conservancy, the N.F.U. and British Rail.

Its purpose would be to receive advice from technical sub-committees on environmental aspects of the mine; 'to integrate the overall requirements into a co-ordinated approach' so that alternative ways could be judged before someone took final and irreversible decisions. He wanted it also to receive progress reports about the mining and to monitor progress, to help communication between the parties on the committee, and also to receive and investigate com-

The village of Wistow, near which pairs of shafts will be sunk.

plaints from the public about the mining.

Much of such a watchdog's work would undeniably make work a little slower for the mining concern. Yet it might also be the price of progress and the proof that talk about participation was not merely empty. The Secretary of State in a recent planning decision for a power station at Inswork Point in the South West of England had already made a committee on similar lines a condition of planning approval.

A further, most useful role the public inquiry played, for all its imperfections, was in letting the local people and others know what to expect and what not to fear. Europe's largest and newest mine was clearly going to need a large labour force. At first, some 2,000 miners were thought to be needed; later this figure was revised to 3,000 and then to 4,000 miners. Such a great migration of people raised fears in the communities from which they might come and the community to which they might go.

West Yorkshire and its older mining communities soon showed they were worried. A free-for-all recruitment campaign for the Selby mine could mean a population drain for West Yorkshire and pit closures there, said West Yorkshire Strategic Planner, Martin Bradshaw. There would be 'destruction of community life.' West Yorkshire pits were anyway undermanned and border-line pits might find their working-lives reduced. 'Where a pit would in the near future reach exhaustion,' he said, 'that pit must close in any event. But where pits still have profitable lives, albeit short ones, as is the case in parts of the western coalfield, it would be unfortunate if the effect of opening Selby was to denude those pits of experienced mineworkers and precipitate closure.'

Before the public inquiry had opened, the Coal Board was well aware of these fears and it was able to give written undertakings to the West Yorkshire Metropolitan County Council. The Board said it recognised that if there was a vast migration of labour from West Yorkshire, it would damage the communities there. It said it must recruit for Selby, but it had no intention of closing other collieries simply to man Selby. The local authority and the Coal Board were agreed on this, and also that as far as possible they would encourage miners to commute to Selby from places like Knottingley, Castleford and Pontefract, where the distances are not too great. Some miners, though, were going to need houses in the Selby area for their families; commuter-miners would never be plentiful enough alone. Again, both the local authorities and the Coal Board were agreed that one thing was to be avoided at all costs: a mining township, or even a number of specific mining villages, little ghettoes for miners. Both planners and mining men have learnt, at a cost, the dangers of building up villages or small towns totally dependent on one form of employment, lacking in diversity, and, in an area like Selby, alien among the local villages that already exist.

The people and streets of Selby.

The North East of England has more than enough mining communities that have lived a half-life because now their pit has closed. One village is Cambois, a settlement on the Northumberland coast that shelters from the North Sea behind sand dunes. I visited it in the spring of 1972, and I found the community engaged in a hopeless struggle against the county planners who, reasonably enough, had decided not to allow any new housing in the village because it was a dying community anyway with dreadful housing. 'The two-up and two-down houses are mean, bathless and without hot water,' I wrote then. 'Their ovens are antique, their chimneys smoke and their lavatories – known as netties – stand like guardsmen in a double-row down the centre of the street.' I found that the pit had gone, the Co-op had gone, the school had gone, the barber had gone, and even the vicar had gone. Only the villagers, growing old, were left behind.

No one wants another Cambois to reappear perhaps fifty years hence in the Vale of York. Provisional plans the local authorities are pondering include scattering the miners around the villages and Selby itself. It won't be easy: Selby District Council planner Roy Smith pointed out that four of the five shafts, down which the miners will go to work, are located to the east of the Ouse where the villages

have in the past experienced very little development. Any large development in these little villages could clearly do much damage to them. Selby's planning committee was of a mind to put 1,000 new houses in Selby, Barlby and Osgodby; 200 houses in Carlton; 200 in Tadcaster; 150 in Eggborough; 150 in Hambleton; 150 in Cliffe; 100 in South Milford and 50 houses each in a further nine villages. This is a far cry from the miners-only villages of the old coalfields.

By the time the public inquiry ended after thirty-eight working days, it would be untrue to say that *all* the local people welcomed the coming of the mine, or that the Coal Board had been converted to *all* the opinions of their opponents. What was happening, however, was a coming-together of the mine's proponents and its objectors; a practical spirit was abroad that the mine *was* coming and that now the best efforts should be directed towards making it a good mine, in a good community, and maximising the benefits for all. Perhaps the greatest challenge to all concerned is to see whether this aim can be lived up to, to see whether a huge mine can be inserted into the Vale of York without bringing with it, in later years, a gradual decline of the landscape and the communities into the semi-industrial mess of random development and dereliction that exists in so much of Britain. This is a challenge as much to the Government, which can take steps to permit the mine but not the mass of less desirable development that might try to follow after. The people of the Vale of York seem to be of a mettle to keep their rulers up to this mark.

On 1 April 1976, the Secretary of State for the Environment, Anthony Crosland, finally delivered his decision on the inspector's report of the public inquiry: he granted the Coal Board planning permission. As almost everyone expected, he felt the richness of coal under the 110 square miles around Selby was a national asset too valuable to ignore. He shared the inspector's view that there was no sufficient justification for seeking to delay the project or limit its overall scope: 'The development would have significant social and economic repercussions, and environmentally its effects would be widespread and some of its consequences adverse. But the Secretary of State is satisfied that, with suitable safeguards, it can be carried out without undue detriment to the affected communities, and in particular, consistently with adequate flood and drainage management, the continuance of agricultural production and the maintenance of public services, and with the planned development of Selby and district, including the settlement of miners and their families.' It is these assertions that the people of the Selby area will have to strive to see are properly met.

Anthony Crosland, however, set out a number of conditions in his planning permission that should help them. One of the most general and interesting conditions was his insistence that the gradual development of the mine should be constantly monitored. To this end, he ordered the setting up of an environmental consultative committee to advise on such problems as dust and noise. He required the N.C.B. to make arrangements with the approval of the County Council to monitor the subsidence and the traffic. And he hinted that the planning authorities should set up some body to consider the overall development of the mine.

On top of that, he laid down a host of other conditions. Among these were: pillars of coal must be left under central Selby, and mining around the town must be restricted to limit subsidence (and that goes for Cawood, too); no mining within a specified distance of River Derwent or the tile works near Escrick without the County Council's approval; subsidence must be limited to 0.99 metres unless the County Council and the Water Authority agree to more.

He did not, however, lay down a condition to maintain, as many Selby people hoped, the Selby–York section of the main East Coast railway line. He felt it could only be kept open by leaving unworked 'an unacceptable amount of coal incompatible with the project'.

Inescapably, Mr Crosland's decisions pleased many but angered not a few. It was never the case that such a huge project as the Selby mine could be carried through without disrupting and disturbing some people's lives or habits. But now, in the Vale of York, the key questions are not whether the mine will come and whether miners will be the new migrants to the Vale and whether the land will subside, but *how* these coming events will be managed.

6. New Uses for Coal

W. T. Gunston

Most of us have a superficial idea of what kind of stuff coal is, though it is no longer common in our homes. We also recall diverse lists of things made from coal – fertilizers, explosives, plastics, inks, paints, perfumes and aspirin – which blur the image of coal as something black and inflexible and, if we care to think, show that the product that emerges at the pit-head is merely a starting point. For generations visionaries have made pleas for coal to be used in more thoughtful ways, and since 1945 research into coal utilization has grown increasingly swiftly until we now stand on the brink of the most exciting period in man's energy history. Only now, it seems, can we view the scene with anything like unfettered vision. Ahead lies a vista devoid of any particular energy crises, provided man loses no time in making more careful use of the precious carbon and hydrogen at his disposal.

There are still many who have no inkling of what can be done to strip down molecules of these two vital elements and synthesize others. Before getting down to the details of 'new uses for coal' it must be emphasized that we have available to us a virtually continuous spectrum of fossil fuels. They are differentiated by hydrogen content. At one extreme is anthracite, a hard solid with little hydrogen and almost all carbon. At the other extreme is methane (almost synonymous with natural gas) which is a carbon atom linked to four atoms of hydrogen, forming a volatile, colourless gas. In between are found every kind of coal, petroleum and all its products and a wealth of other solid, liquid and gaseous fuels and chemical feedstocks. The point is: we can turn one into another. Sometimes the conversion yields energy, but usually it consumes it. What we have to learn is what is today blindingly obvious to only a few: that we are at present foolishly wasting and misusing almost every one of Earth's hydrocarbon resources. Great individual stature is called for – rather urgently – to take correct decisions of gigantic magnitude in technical, political, social and financial fields that will affect all the world's population.

We have already taken the first and most obvious steps. Hardly anywhere in the world are lumps of coal merely thrown on to an old-fashioned open grate. Today the pit-head is recognised as standing at the beginning of a great highway that branches off into the new pathways that this chapter is about. Some of the pathways link with others, turn back on themselves or tie up with totally different families of energy or materials. (For example, there is much to be said for using nuclear heat to evaporate water in a steam boiler and then burning coal at higher temperature to superheat the steam to the desired

Pre-moulded heat-shrinkable gloves are one example of the many by-products of coal tar. They seal cable ends, insulate electric joints and protect exposed or delicate equipment.

Left: Coke being pushed out of an oven before quenching.

quality for a turbogenerator.) But first there is one large use of coal that stands to some degree apart from national energy supplies and raw materials, and which appears likely to continue for as far ahead as can be seen. This is the production of metals, especially steel.

Although there are other methods of providing energy for iron- and steelmaking, the traditional means is the blastfurnace. This has long been associated with the provision of good high-quality coking coal, to make big lumps of coke strong enough when white hot to support a column of coke, ore and other material more than fifty feet high. Such coal is in short supply, but recent research has resulted in a technique of mixing more common varieties, especially in the form of briquettes, to form ideal coke in the blastfurnace. It is fortuitous that coke in the blastfurnace can simultaneously support the contents, provide carbon for heating and carbon monoxide for reducing the ore to the metal, while remaining porous to the upward flow of hot gas. Coal's position as the main reductant for metal ores all over the world does not appear likely to change. At the same time it must be admitted that what goes on in a blastfurnace is still imperfectly understood, because research is for obvious reasons difficult. Direct injection of pulverised coal and/or reducing gases made from coal appear to be routes to reduced overall costs, and even the 'char' residue from the formed-coke can be put to use.

Char can be 'gasified', as described below, but a more immediate use is in a fluidised bed combustor (F.B.C.). The technology of fluidised beds is vast and originally had nothing to do with coal. Such a bed is a container of finely divided material through which gas is blown from the bottom. At a certain velocity the particles in the bed become levitated by the flow; a bed of flour would clearly work with much lower velocity than one of sand. In operation the bed behaves like a liquid, or rather a boiling liquid because bubbles of gas rise through it and the surface is diffuse. During the past decade it has

become increasingly self-evident that, if any solid fuel is to be burned, the F.B.C. is the way to do it. An F.B.C. is basically a fluidised bed of refractory and inert particles into which fuel is injected or dropped. The advantages are very widespread.

Today the most efficient established way of burning solid fossil fuel is in a large P.F. (pulverised fuel) furnace. The boilers of a major power station are each as large as a city office block. They combust the fuel in flames about 100 feet high, each flame having essentially streamline flow and intensely high temperature to transfer the heat to the boiler tubes mainly by radiation. Molten slag or clinker continuously solidifies, often tightly adhering to solid surfaces. In contrast, combustion in an F.B.C. takes place at a mere 800–900°C, well below slag-melting temperature and in the exact band of temperature at which dangerous sulphur dioxide can readily be 'fixed' by basic reaction with limestone or dolomite. At this relatively cool temperature no significant oxides of nitrogen are formed, nor are alkali metals volatilised. Yet, despite the lower temperature, combustion in a

The steam tubes (below) in an early fluidised bed burner at the Coal Research Establishment, Stoke Orchard, correspond to the steam tubes shown in the diagram (right) explaining the principle of fluidised bed combustion.

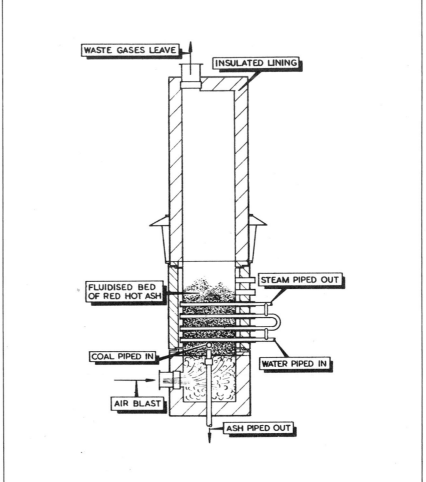

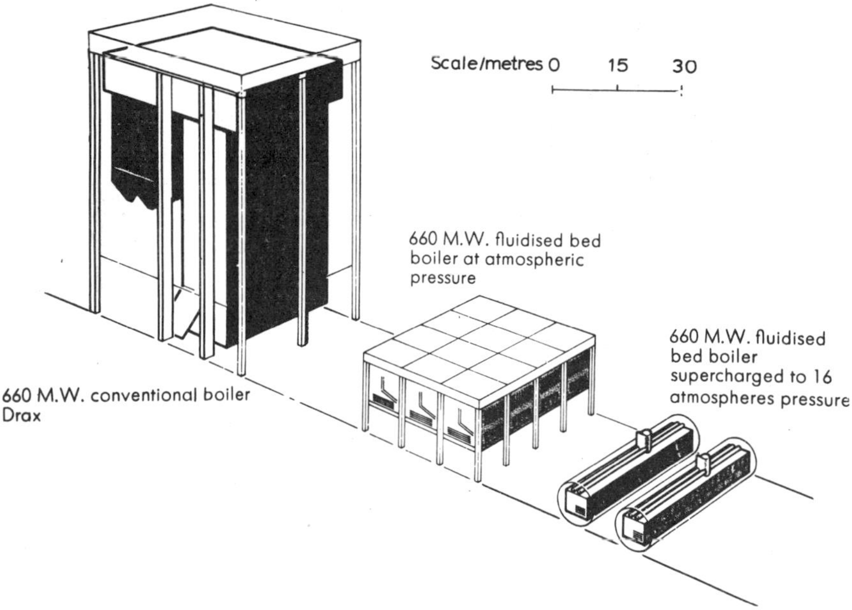

Left: The relative volumes of a conventional boiler used in a power station and projected fluidised bed boilers.

[Diagram labels: 660 M.W. conventional boiler Drax; 660 M.W. fluidised bed boiler at atmospheric pressure; 660 M.W. fluidised bed boiler supercharged to 16 atmospheres pressure; Scale/metres 0 15 30]

fluidised bed is far more intense, measured in terms of heat released per unit volume per unit time. This is because each fuel particle is surrounded by its own constantly changing atmosphere, and heat transfer is primarily not radiative but by solid/solid contact, which makes the process about twice as effective. This means an F.B.C. can be significantly smaller than a conventional boiler, with capital savings of 12–20 per cent (rather more if sulphur emissions are controlled) and operating savings of 10–16 per cent.

There are many other advantages, but the F.B.C. can be further and dramatically improved by operating it at high pressure – say, 16 times atmospheric. This cuts the size to a fraction of that needed for atmospheric-pressure operation, and opens up many fresh possibilities of which not least is that the filtered flue gas can be expanded to atmosphere through a large gas turbine. The N.C.B. has led intense world research into this exciting field, and calculations show extremely attractive possibilities for replacing today's enormous electricity generating plant – at least for medium-load infilling – by compact 100 MW sets, each no larger than a typical room, combining steam and flue-gas turbines and affording great benefits in higher efficiency and reduced emissions. Moreover, such installations are insensitive to the quality of fuel. They could with advantage be supplied with coal from seams containing a high proportion of dirt, or with colliery 'tailings', the incombustible matter simply helping form the F.B.C. bed along with the ash. The latter automatically grades itself into fine dust extracted by cyclones in the flues and larger particles removed from the bed itself – all, of course, useful by-products.

Fluidised beds appear extremely promising in the conversion of coal into gaseous and liquid fuels. Gasification of coal is essentially a hydrogen-adding process, with historic origins in the Lurgi technique

Above left: A fluidised bed slurry burner built at the Coal Research Establishment. It at last makes possible easy disposal of this unsightly waste.
Above right: A fluidised combustion steam boiler of 14MW in use at Babcock and Wilcox's Renfrew Works. The bed is 10ft square, will burn either solid or liquid fuels and retain sulphur during combustion.
Right: The Development Centre of the British Gas Corporation at Westfield, Fife where methods of converting coal into substitute natural gas (SNG) were developed in 1975.

developed over forty years ago. This was a fixed bed process, in which the coal was broken down under pressure by steam and oxygen to yield a synthesis gas – mainly carbon monoxide and hydrogen – which can then be combined in the presence of a catalyst to yield methane. Today there are experimental plants with entrainment beds, in which dispersed-phase particles are blown along the bed by the gas flow. Later fluidised bed gasifiers promise even better efficiency and economy in capital and operating costs.

Out of such gasification plant can come a range of convenient fuels from 'syncrude' through methanol, to hydrogen and methane (substitute natural gas, or S.N.G.). In Britain there is understandably some reluctance to understand why natural gas should be manufactured, because the fact that North Sea supplies may diminish very sharply after 1981 has not been widely publicised. But in the United States the impending shortage of this convenient and clean-burning fuel has long been understood and in the next decade U.S. production of S.N.G. will rise to several hundred million tons per year. The world's leading coal gasification research centre – at Westfield, Scotland, operated by British Gas with the benefit of forty years of experience – has been supported in new work solely by a large group of U.S. companies and organizations. British interest is confined to the few experts who can see the future.

It can be shown that, even allowing for the lower efficiency of gas appliances, it is far more efficient to convert coal into S.N.G. than into electricity. With present technology the proportion of energy in the coal that reaches the consumer by electrical means is 27 to 32 per cent, whereas conversion to S.N.G. approximately doubles the fraction to 56 per cent. This is in part because gas is so very much cheaper to transport than electricity, and even allowing for the fact that S.N.G. from coal will be at least twice as costly as true natural gas is at present, it seems certain that S.N.G. will be the biggest single fuel in most countries by the beginning of the new century. Probably there will be local regions, extending to about 12 miles around each gasification plant, within which it will be cheaper to send out the low-BTU synthesis gas (so-called 'lean' gas), which is cheaper to make but costs more to transport by pipeline.

It is fundamentally important to note that coals contain many useful ring or chain hydrocarbons – ethylene, benzene, toluene and many other groups and families – which in previous coal conversion processes were largely destroyed. As building such molecules consumes energy it is clearly preferable to retain them, and much effort has been devoted to pyrolysis and solvent-extraction or hydrogenation processes that do so. Pyrolysis is thermal decomposition at moderate temperature in the absence of air or steam. As no hydrogen is added, a solid char is left which can be converted into gas; partly because of this the U.S. Cogas Development Corporation is sponsoring the N.C.B. research on this process at Leatherhead. Unlike old-fashioned coke ovens, which continuously manufactured large amounts of valuable benzene, methane and other products and let them im-

The N.C.B. continues its research into the basic characteristics of coal. This photograph taken in the 1950s, shows an investigator at the Coal Research Establishment determining the nitrogen content of coal.

mediately recombine, modern processes aim to separate and quench the volatile product quickly. Put another way, if coal is to be burned, it pays to see what valuable material can be extracted relatively cheaply beforehand. There is a wealth of possibilities. Most coals contain huge carbon molecules combined with hydrogen, plus some oxygen, nitrogen, sulphur and traces of other elements. Though any pyrolysis process can be tuned, mainly by adjusting temperature to give the best yield, it is rather insensitive to the type of coal over a wide

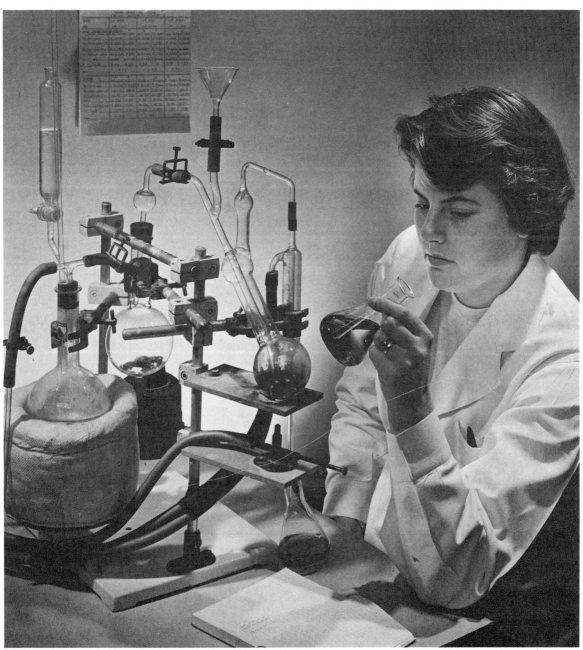

range of medium-volatile types.

An alternative method – which, like gasification and pyrolysis, can be combined with others in a hybrid process – is direct hydrogenation or solvent extraction. This can assume various forms but the N.C.B. Coal Research Establishment at Stoke Orchard dissolves a slurry of ground coal in anthracene oil, itself a coal derivative, at about 400°C. As the solvent donates hydrogen it becomes depleted, and so is not only recovered by distillation but also restored by adding hydrogen either to the solvent or to the vessel. The digest, the coal solution in the reaction vessel, is subjected to further treatment and filtration to yield a range of products.

A variation of the solvent-extraction process is to use liquids such as toluene or even water at just above the critical temperature (the temperature at which the vapour can no longer be converted back to liquid by pressure alone). Supercritical liquids have a peculiarly great power of dissolving solids, as happens in natural reactions between steam and rock in the Earth's crust. In coal solution the supercritical method has the advantage of giving more freedom to choose temperatures and pressures exactly tailored to the recovery of the highest fraction of the best molecules. The brown liquid that results can be distilled directly, in the manner of crude petroleum, or can be subjected to various catalytic hydrogenation processes before allowing the desired fractions to condense or precipitate out on lowering the pressure and temperature. This, the most elegant method yet devised for breaking coal molecules into valuable pieces, can be used with large lumps of coal without need for pulverising or grinding, and also yields a low-grade carbon residue suitable for combustion or gasification.

In the United States there has been much interest in making solvent-refined coal (S.R.C.), a heavy, oily liquid used merely as a fuel. Though this could serve as a replacement for increasingly scarce petroleum, and burns more cleanly than the sulphurous American coals, it is – in the N.C.B.'s opinion – bad economics to build huge plants merely to make something to burn. The overriding urge behind the British research into better coal conversion processes has been to obtain scarce and valuable raw material. Today the organic chemical industry is based on petroleum, yet chemical feedstocks represent a mere 10 per cent of the total and have the status of secondary products. In thirty years there is no doubt that there will have been a complete change. Feedstocks for plastics manufacture alone have been increasing at over 8 per cent (compound) annually, much faster than the growth in human energy supplies, and by the start of the new century these will be primary products. Depending on the rate at which the oil wells are run down, organic chemicals will increasingly have to come from coal, without having to go back to the basic building blocks of carbon monoxide and hydrogen. Hence the enormous British drive to perfect processes for the conversion of coal molecules.

To a considerable degree this is a reversion to the situation that

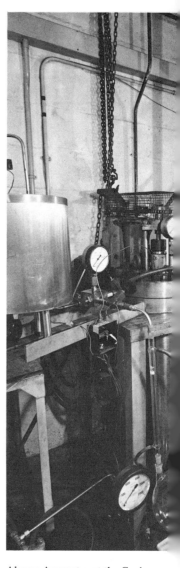

Above: Apparatus at the Coal Research Establishment for dissolving coal using gases under supercritical conditions of temperature and pressure (supercritical extraction).

Above right: A flow sheet showing the cycle of solvent extraction of coal and the products possible at each stage.

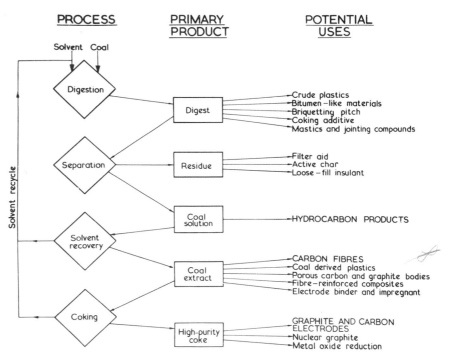

existed in the nineteenth and early twentieth centuries, when coal was by far the most important source of feedstocks for organic chemicals. This was not deliberate policy but merely resulted from the large-scale manufacture of town gas, which automatically created various tars, liquors, benzole and other products. Around 1950 the massive expansion of the chemical demand, combined with the phasing-out of the manufacture of town gas, threw virtually the whole burden on to the petroleum industry, which is today being depleted very rapidly (witness the fact that the amount of petroleum consumed in 1960–69 was roughly the same as the total amount consumed from the date of the first oil well, 1852, to 1960). Most observers agree that growth in petroleum output must slacken around 1990, and become negative in about 2000. All present predictions show the curve of petroleum production falling very rapidly by 2020, while world population and energy demands continue to rise ever more steeply. It is an accident of history that this book appears at exactly the time when petroleum is, like a supernova, rushing to its brief peak of brilliance.

If there were no revolution in the source of chemical feedstocks, the entire world output of petroleum would be needed for this purpose alone by 2020, leaving none for what is today's dominant need, energy for homes, industry and transport. So increasingly we may expect to see dramatic and rather sudden changes in the ways in which fuels are used, and these will revolutionise man's energy supply picture. With luck it will be a quiet and painless revolution that will not be superficially evident – if we start it now.

The first major change, which is fortunately beginning already, is for petroleum and natural gas to be recognised as extremely valuable and limited materials which we can no longer afford merely to burn for base-load power generation. Increasingly they will rise in price to something closer to their replacement cost, while being reserved in application strictly for uses demanding portability, clean burning and as a source of feedstocks. Ultimately coal will also have to follow petroleum and gas up the 'value ladder', and the timing depends very much on progress made with nuclear reactors of improved types. The Earth's coal reserves are so gigantic that resource problems are nothing like so imminent as they are with petroleum, but ultimately coal, too, must cease to be used for mere base-load combustion. Today Britain, for example, can still afford to plan new coal-fired power stations (though of radically new design, with F.B.C. and gas turbines as well as steam), but in the fairly near future all base-load heat will derive from nuclear reactors, with combustion in the classical sense used only for intermittent or portable energy demands.

Earlier the suggestion was made that nuclear heat is ideal for steam generation, while high-temperature combustion of coal and other fossil fuel could best be utilised for improving the quality of the steam by superheating. This alliance on one site between nuclear and fossil fuels is merely one facet of a process that has been simmering in the background for generations but has rarely been able to emerge into the limelight (often through lack of a sufficiently broad-minded sponsor). The process can best be called energy integration; it is seen already in a primitive way in district heating schemes which use surplus low-temperature heat from power stations that would otherwise be wasted (we often forget that the heat rejected to the atmosphere and/or to lakes and rivers by power stations vastly exceeds the amount sent out as electricity).

In the context of coal it is only common sense to do as many complementary things as possible on one site, and tie the processes together to maximise the quantity of energy and raw materials sent out. The natural word for such a coal-processing complex is coalplex, and the variety of possible forms is open-ended. Most, however, involve at least one pressurized F.B.C., at least one convertor (solvent extraction, gasification, hydrogenation or similar process), and yield a mix of raw materials, heat and electricity. The main objective with a coalplex is to minimise losses which would inevitably be suffered were the processes to function separately. In an extremely large installation there would, additionally, be some advantages in scale. On the other hand it would be mistaken to create a giant edifice that was inflexible and essentially unchangeable in its operation. For example, the C.O.G. refinery proposed in the United States would deliver S.N.G. and syncrude; it would not be possible to stop producing one without also stopping the other.

A more flexible coalplex, of a particularly simple kind, would be used to make S.N.G., plus, in the peak-demand daytime periods, electricity. What makes electricity more difficult than other energy forms is that, with present technology, it is grossly uneconomic to store. On a national scale it has to be generated at a rate that matches

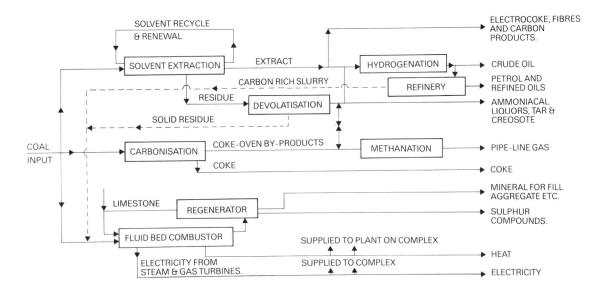

A possible outline scheme for a coalplex.

demand, and in practice this means providing a constant base-load (even this varies greatly on a seasonal basis) and supplementing it with other capacity on a diurnal basis. Until now this extra capacity has been unable to do anything in off-peak periods except, in some cases, attempt to store energy by pumping water to a high reservoir. With the proposed coalplex, the basic energy stream can be switched entirely to S.N.G. production and high-pressure storage, with substantial gains in overall efficiency. In addition, of course, such an integrated plant would gain in overall thermal efficiency and in reduced transport and administrative costs. Once each individual process had been optimised the objective would be to integrate them in such a way that electricity, steam, gas, syncrude, ash and chemical feedstocks could each be varied without crippling the efficiency of the plant as a whole or the output of any other commodity. This is by no means simple to do, but extremely good results can be attained which are greatly in advance of present practice with separated plant.

There are, of course, fundamental limits and rules. Coal is easy to store but not very cheap to transport, so on this basis it would be logical to site coalplexes near the pits. Electricity is expensive to transport and virtually impossible to store. Gas, on the other hand, is very cheap indeed to transport, and fairly cheap to store. Unit costs of any process are to some degree dependent on load factor, and it is undesirable to have a major plant run for only a few hours per week. By no means least, each atom of carbon cannot be used more than once.

It is equally fundamental that conversion costs money. Today's vast tonnages of coal burned directly in P.F. boilers do at least avoid these costs, and also release virtually all the energy that can be released by simple combination with oxygen. Adding hydrogen to coal to make gasoline (for example) involves both capital and operating costs, as well as a loss in total fuel energy of around 40 per cent, so that the final product is unlikely to cost less than 2.5 times as much (though modest improvements can be foreseen). But such questions are surely academic. In the short term, availability of petroleum products from coal will provide a constant assurance against excessive price-rises by the oil producers. Though there is a surprisingly widespread inertia and complacency in Europe, the United States has made it clear that large-scale hydrogenation of coal is a distinct possibility, and such states as Israel and, especially, Japan, regard such a source of liquid fuels as very likely within another decade. In the longer term, of course, there will be no other source.

This raises further issues, to which politicians and the general public ought, perhaps, to devote a little attention. The world's petroleum industry cannot be allowed to go on increasing output until suddenly we wake up one morning and find it has all gone. In practice, of course, it would not be like that; new reserves are being discovered and with improving technology it becomes economic to work poorer fields, ultimately moving into the vast tar sands and other difficult sources. But none of this invalidates the central issue that petroleum output is certain, in so far as anything can be said to be certain, to decline from the first decade of the next century onwards, while human demand rises at ever-greater rates. The next twenty-five years are clearly crucial to human life. We must use them to the full to find the best answers to the obvious problems of the twenty-first century, when natural petroleum and natural gas gradually become part of history.

Though the British government naturally could not be expected to appreciate this, we ought not today to use a single atom of carbon in petroleum or natural gas unless we have to. It is not insulting the intelligence to point out that the faster we use these fuels the sooner they will run out. And there is a further major factor which the politicians and public ought to consider. Every bit of oil or gas we use has a replacement cost between two and three times as high as the price at present charged for it (by 'cost' is meant the price charged by the producer, and not that charged to the consumer after the addition of several hundred per cent duty). Coal could make gasoline (petrol) at a price not greatly different from that at present charged the consumer in many countries, but what would be the policy of governments regarding the duty payable, which is a major source of public revenue? Even more sobering is the thought that today most natural gas is untaxed, and is sold at a price roughly one-third as high as the price of S.N.G. in the future. There may be a case for examining this situation and perhaps progressively increasing the price of natural gas over the next several decades up to the level of the price of S.N.G. It is too much to hope that the revenue gained might be applied to reducing the capital and conversion costs of coal when this becomes the sole source of fossil fuels.

Few would care to make any detailed prediction of man's energy sources and uses in the next century, beyond the self-evident fact that coal and nuclear power will predominate. As far as can be seen, nuclear power will continue to be confined to the production of heat in fixed stations. It does not appear of itself to be portable, nor to have relevance to chemical feedstocks. Coal, therefore, will ultimately have to provide not only the carbon atoms for plastics and countless other materials but also all the liquid and gaseous fuels other than hydrogen itself. The extent to which hydrogen, presumably from water, can support an energy economy is too big and separate an issue to describe here. What is beyond dispute is that the Earth's coal resources are enormous, and with improved technology a very much greater fraction will be recoverable than at present. Though it may be premature to discuss it now, the prospects for mining in the more distant future are as exciting and encouraging as any part of man's history.

It has happened by chance that coal has for many years been on the defensive. It still has an image of black griminess, of foul working conditions, of all that was worst in the Industrial Revolution. The very fact it has been around so long inevitably suggests at least obsolescence, and its labour-intensive nature has hit it hard indeed in a world where labour is no longer cheap. On top of all this, the rivals of nuclear power and petroleum have sprung into prominence, with images of newness and growth and, whether real or imagined, lower costs. This has led to a situation which, in Western Europe (and Western Europe alone), has been compounded by deliberate rundown in coal output and, until two years ago, the open portrayal of coal as a declining industry. An observer in the next century would find all these things very hard to understand.

Far from being on the defensive, coal is belatedly recognizing its true worth. Far from being obsolescent, coal is fast recovering its position as man's chief and most versatile energy source. There is nothing petroleum and gas can do that cannot be done by coal, at greater cost. The fact that we can enjoy artificially cheap oil and gas for a few decades affects only the present population. The whole history of fossil liquid and gaseous fuels will be but a brief moment in the lifetime of the human race, and today's children will see its decline. When that occurs, coal output will still be rising – faster even than it is today – and an ever greater proportion of the coal will be used as a source of carbon and hydrocarbon compounds. None of it will be merely burned. At the same time, for most of the immediate future coal will still serve primarily as a source of heat. It is a flexible source, giving heat at moderate temperature or high, rapidly or slowly, intermittently or continuously and available at short notice. Even more significant, it is man's chief and long-lasting resource of carbon which, with sunlight and oxygen, he needs in vast amounts in order to survive.

Index

References in italics refer to captions to illustrations.

Accidents 50
Ackton Hall Colliery 97
A.F.C. *see* Armoured face conveyor
Allied Mills Ltd 119
Ancient mining 32–5, *33*, *34*, *35*, *36*
Anderson Mavor ranging-drum shearer *83*
Armoured face (or flexible) conveyor 48, 52, *55*
Ashington Colliery 34
Athabasca tar sands 11
Automation and remote control 93

Babcock and Wilcox fluidised bed boiler *133*
Barlby 127
Barlow No. 2 borehole 98, 104
Beeston Seam 98
Bell pit 32, *33*
Betws drift *22*
Bevercotes Colliery *70*
Bishop's Wood, Wistow 124
Blastfurnace 130
Blasting *38*
Blelloch, James 123
BOCM Silcock Ltd 108, 119
Bradford 97
Bradshaw, Martin 126
Bramley, Makin 121
British Gas Corporation 134, *132*
British Rail 116, 124
British Waterways 116

Camblesforth borehole 98
Cambois 126
Cannock Chase *41*
Carboniferous swamp *30*
Carlton 127
Castleford 97, 126
Cawood Common borehole 100, 108, 111, *98*
Central Electricity Generating Board 22, 96, 112, 122
Chadwick, Dr Michael 124
Chainless haulage *60*
Cliffe 127
Club of Rome 10, 11
Coal
 anthracite 32, 129
 bituminous 32
 Bronze Age use of 29
 Carboniferous 29
 chemical feedstock 129, 137, 141
 Coalplex 7, 138, *139*
 convertors 138
 direct hydrogenation 136
 electricity generation 138–140
 face *see* Coalface
 fossil plant remains 29
 gasification 132, 136
 hydrogenation 129, 136, 140
 inland consumption 24, 26
 Lurgi process for gasifying 132
 methanol 134
 organic chemicals 137
 plastics 136
 pulverised fuel boilers 140
 pyrolysis 134, 135
 raw material, not only a fuel 136
 resources 12, 18, 19, 20, 27, 28, 138
 semi-bituminous 32

S.N.G. *see* substitute natural gas
solvent extraction of 136
stocks of 52
strip mining in USA 27
sub-bituminous 32
substitute natural gas from 134, 138
supercritical extraction of 136
syncrude 134, 138
synthesis gas from 134
vegetable origin 29, 31
see also Early mechanisation underground *and* Mechanisation
Coalcutter 4, *see also* Early mechanisation underground *and* Mechanisation
Coalface 42 *see also* Mechanisation; prop-free-front face
Coalface lighting *50*
Coalmining 56–57
 ancient mining in China 29
 ancient mining in Britain 13, 32, 33, 42
 belt conveyors 43, 45
 Channel Tunnel (1882) 86
 dinting 64
 early hoisting and pumping methods 34, 35, 38
 early underground mechanisation 46–8
 electrical equipment 46–7, 61
 Inspectorate 52
 jigger conveyors 43
 longwall mining 47, 56, 57, *53*
 packing 81
 pillar extraction 38
 pumped packing 81
 productivity 52, 61
 retreat mining 63
 ripping 64
 ripping lip *64*, *81*, *84*
 roof control 48, 50, 54
 room and pillar mining 38, 43
 scraper-chain conveyor 43, 47
 shift working, three-shift working 52
 see also Cutter-loaders, Mechanisation, Underground transport
Coalplex 7, 138, *139*
Coal Research Establishment 7, *75*, *135*
 fluidised bed burning of coal 130–32, 138, *131*, *132*, *133*
 heatshrinking *129*
 reflectance microscope for determining coal rank *31*
 solvent extraction of coal *137*
 supercritical extraction of coal 136, *136*
Cochrane and Sons 120
Cogas Development Corporation 134
Coke pushing *130*
Compressed air transport *44*
Computer Power Ltd 4
Crosland, Anthony 128
Cublington 101
Cutter-loaders 52, 59
 Anderton shearer 48, 81, *49*, *60*
 bi-directional 83
 double-jib *49*
 ranging-drum 83
 plough 81
 trepanner 48, 81, 83, *83*
 see also Early mechanisation underground *and* Mechanisation

Danish Bacon Co. Ltd 119
Dawdon Colliery *69*
Derwent, River 104
Dickie, Alan 121, 122

Doncaster 98, 108
Drake, Col. Edwin 13
Drax Power Station 101, 108, *103*
Dust 87, 89

Early mechanisation underground 46–8
 coalcutters 46
 Dosco 48
 Duckbill loader 47
 Gloster getter 48
 Huwood loader 47
 Joy loader 47, 48
 Lee-Norse miner 48
 Logan slab-cutter 48
 Shelton loader 47
 shuttlecar 47
 Uskside miner 48
Eggborough 127
Eggborough Power Station 101, 108
Energy integration 138
Environment, Department of the 114, 120, 128
European Community 18
Explosion of gas 89
Ezra, Sir Derek 112, 113

Featherstone 97
Ferrybridge Power Station 101
Flameproofness 87–8
Fluidised bed burning of coal 7, 130–32, 138, *131*, *132*, *133*
 pressurised 132
Formed coke 130
Forrest, W. 100, 105, 107, 112
Fryston Colliery 97
Fuels
 comparison 139
 conversion 129, 138–140
 crisis of (1973) 6, 8, 100, 111
 input to electricity and gas 25
 marketing and reserves of 6, 10, 11, 13, 16, *12*
 usage and living standards 8–11
 waste of 11, 129

Gas (methane) 87, 129
 substitute natural gas 134, 138
Glasshoughton Colliery 97
Goossens, R. F. 98, 108
Griffiths, John 116

Handcutting of coal *41*
Heatshrink *129*
Hemingbrough borehole 98
Henry VIII 97
History of mining *see* Ancient mining *and* Early mechanisation underground

In-seam miner *63*, *69*
Inswork Point Power Station 125
International Energy Agency 7, 9, 28
Intrinsic safety 87, 88

Jevons, Stanley 10
John E. Sturge Ltd 120
John Rostron and Sons Ltd 120

Kelfield Ridge borehole 98
Kellingley Colliery 100, 104, 113, *100*
Kellington borehole 98
Kent, Reverend J. A. P. 111, 118
Kenyon, R. Cooper 120
Knottingley 126

Leatherhead 134
Leeds 97
Leland, John 97
Londesborough, Earl of 98
Longannet Mine 113, *22*, *90*
Long Lane, Wistow 124

Malthusians 11
Manless mining 93
Maplin Sands 101
Marco Polo 29
Marsh, Gentian 122
Mechanisation
  accidents 50
  armoured face (or flexible) conveyor (A.F.C.) 48, 52, 55, *55*
  caving 48, 54
  chainless haulage 59
  continuous mining 54
  cutter-loaders 52, 59
    see also Early mechanisation underground
  dinting 64, 84, 86
  Dosco dint-header *86*
  Dosco roadheader 84, *85*
  face room 54
  full-face tunnelling machines 84, 86, 87, 95, *73*
  gate-end box 61
  hydraulic props 50, 55, 59
  impact ripper 84, *85*
  longwall face 47
  maingate 61
  packing 63, 64, 81, 84, 86
  powered supports see self-advancing props
  productivity 52, 61
  prop-free-front face 48, 52, 54
  retreat mining 63, 84
  ripping 63, 64, 84
  ROLF 50
  roof control 48, 50, 54
  rotating-head ripping machine *see above* Dosco roadheader
  self-advancing props (powered supports) 48, 52, 55, *54*, 58
  stables and stable elimination 59
  stage loader 61, 84
  tunnel drivages 63, see also above ripping *and* full-face tunnelling machines
Methane see Gas
Methane drainage 87, 90
Mine rescue 69
Mini-computer on mine surface *64*
Mining Research and Development Establishment
  and Associated Portland Cement Manufacturers 96
  automatic firedamp detector 96, *96*
  automation 93
  Campacker *86*
  and Central Electricity Generating Board 96
  chainless haulage *60*
  coalface investigations 96
  coal preparation investigations 96
  comprehensive monitoring investigations 96
  dust reduction 90
  in-seam miner *63*, *69*
  integrated circuits, use of 93
  irrigated dust filter *89*
  nucleonic shearer 93, 94
  underground environment investigations 95

Vorsyl separator *96*
Modern colliery *53*
Monitoring the underground environment 94

Naburn borehole 104
National Coal Board 6, 7, 20, 21, 52, 87, 112, 113, 124 *and* passim
  drilling for coal 22, 50
  geophysical (seismic reflection) surveying 50
  see also Coal mining; Coal Research Establishment; Mechanisation; Mining Research and Development Establishment; Selby
National Farmer's Union 116, 121, 124
Nature Conservancy 116, 124
N.C.B. see National Coal Board
N.C.B. – Thyssen tunneller see Mechanisation, full-face tunnelling machines
North Sea gas 134
North Sea oil 111
North Yorkshire coalfield 97
North Yorkshire County Council 116, 120
Nuclear Energy Agency of OECD 27
Nuclear power 17, 141
  fast breeder reactor 27
  National Nuclear Corporation 27
  steam-generating heavy water reactor 27
Nucleonic shearer *93*

Orchard, Eric 116
Osgodby 127
Outhwaite, Francis and John 121

Parkgate seam *70*
Parkside Colliery *53*
Permian limestone 98
Petrochemicals 137
Picking belt *46*
Planning Act, 1968 114
Planning Inquiry Commission 114
Pneumatic transport *44*
Pontefract 97, 126
Potts, Professor E. L. J. 116
Powered supports 48, 52, 55, *54*, 58
  see also Mechanisation
Power shovel *86*
Prince of Wales colliery 97, *66*
Pulverised fuel burner 131

Rank Hovis McDougall flour mills 108, 119
Rank of coal *30*, *31*
Reflectance microscope *31*
Renton, Peter 119
Retreat mining *63*
Rio Tinto Zinc 111, 112
Ripping lip *64*, *81*, *84*
Roff, Peter 114
Rope haulage *92 see also* Underground transport

Sanderson, Fred 112, 113
Saudi Arabia 10
Selby 22, 97–127, *75*, *80*, *103*, *109*
  Abbey 100, 106, 108, 117, 118, *109*
  Barnsley Seam 97, 98, *98*
  Cawood *120*
  Chamber of Trade 114
  coalfaces 105
  commuter-miners 126
  Deighton 123
  District Council 116, 119, 120, 124, 126

  drifts at Gascoigne Wood 101, 104, 107, 114, 122, *66*, *103*, *106*
  Environmental Consultative and Liaison Committee 124
  Escrick 123
    shafts 124
  Gascoigne Wood *see above* drifts at Gascoigne Wood
  heading machines 105
  high technology mining 105
  locomotives 105
  inclined shafts *see above* drifts at Gascoigne Wood
  Newsletter 112, 113
  North Duffield shafts 124
  Ouse, River 100, 108, 110, 121, *109*
  planning application 112
  Protection Club 114
  retreat mining 106
  Riccall 122
    shafts 124
  Skipworth Common 122
  South Milford 127
  Stillingfleet shafts 124
  vertical shafts 104
  winding tower heights 123
  Wistow *125*
    shafts 122, 124, 104
  see also Subsidence
Shotfiring 38
Skipworth Common 122
Smith, Roy 126
Snaking conveyor *see* Armoured face conveyor
S.N.G. *see* Substitute natural gas
Snowdonia National Park 111
South Milford 127
Staffordshire Thick Coal 29
Stansted 101
Steam engine for mining *35*, *37*
Subsidence 100, 106, 116, 117, *115*
  and water table at Selby 117, 121, 122
Substitute natural gas (S.N.G.) 134, 138

Tadcaster 127
Thorganby *109*
Treadmill *34*
Trepanner 48, 81, 83, *83*
Tube bundles 94, *94*
Turnhead Farm, Barlby 121

Underground transport 90
  belt conveyors 90, 92, *91*
  cable-belt conveyor 92
  hydraulic transport 92
  locomotives 45, 92
  manriding 44, 92
  mine cars 44
  pneumatic transport 92
  rope haulage 92
  skilifts 46
United Kingdom fuel consumption *22*, *27*

Vassie, Maurice 123, 124
Ventilating air 87–9
Vorsyl separator 96, *96*

Wakefield 97
Walker, Peter 112
Warren House seam 98
Warwickshire Thick Coal 29

West Yorkshire Metropolitan County
  Council 116, 126
Wharfe, River 100
Whimsey *37*
Whitemoor borehole 98
Women underground *37*

World coal production *12*
World coal resources *13*
Wressle 98

York 98
  Vale of 108, 110, 113, 117, 123, 126, 127

Yorkshire and Humberside Council for the
  Environment 116
Yorkshire Chemicals 119
Yorkshire Evening Press 123
Yorkshire Naturalists' Trust 116
Yorkshire Water Authority 116, 124

Acknowledgements

The publishers are grateful to the following for supplying illustrations:
Aerofilms Limited 109, 120, 122, 125
Beamish, North of England Open Air Museum 38, 40, 41, 43, 46
A & C Black 56
Coal Factors' Society 66
Ray Green 69
Department of Energy Statistical Digest, 1975 with the permission of the Controller of
H.M.S.O. 24, 25, 26
Leila Kooros 78, 79, 80, 103, 110, 118, 126, 127
N.C.B. 12, 21, 23, 30, 31, 33, 35, 41, 44, 49, 50, 53, 54, 58, 60, 62, 64, 67, 68, 69, 70, 71, 72, 74, 75, 76, 77, 81, 82, 83, 85, 86, 87, 88, 89, 91, 92, 93, 94, 95, 96, 99, 100, 102 (from the Ordnance Survey 1:50,000 map with the permission of the Controller of H.M.S.O. (Crown copyright reserved)), 103, 104, 106, 112, 115, 129, 130, 131, 132, 133, 135, 137
Radio Times Hulton Picture Library 34, 36, 37, 38, 42, 45
From an article by D. H. Broadbent in Inglis (ed.), *Energy: From Surplus to Scarcity*
(Applied Science Publication 1974) 139